Plas

Questions and Answers books are available on the following subjects:

Brickwork and Blockwork
Carpentry and Joinery
Painting and Decorating
Plastering
Plumbing
Central Heating
Refrigeration
Automobile Brakes and Braking
Automobile Electrical Systems
Automobile Engines
Automobile Steering and Suspension
Automobile Transmission Systems
Car Body Care and Repair
Diesel Engines
Light Commercial Vehicles
Motor Cycles
Electricity
Electric Motors
Electric Wiring
Amateur Radio
Radio Repair
Radio and Television
Colour Television
Hi-Fi
Electronics
Integrated Circuits
Transistors
Electric Arc Welding
Gas Welding and Cutting
Pipework and Pipe Welding
Lathework
GRP Boat Construction
Steel Boat Construction
Wooden Boat Construction
Yacht and Boat Design

QUESTIONS & ANSWERS

Plastering

Don Stagg
Brian Pegg

Newnes Technical Books

Newnes Technical Books

is an imprint of the Butterworth Group
which has principal offices in
London, Sydney, Toronto, Wellington, Durban and Boston

First published 1978 by Newnes Technical Books
Reprinted 1981

British Library Cataloguing in Publication Data

Stagg, William Dorrall
 Plastering. – (Questions and answers series).
 1. Plastering
 I. Title   II. Pegg, Brian Frederick   III. Series
 693.6   TH8131   78-40657

 ISBN 0-408-00301-4

Typeset by Butterworths Litho Preparation Department
Printed in England by Fakenham Press Limited, Fakenham, Norfolk

CONTENTS

1
Materials   1

2
Solid Plastering   10

3
Plasterboard   35

4
Floor Screeds and Pavings   40

5
Run Plain Mouldings   46

6
Open Moulds   63

7
Casting   70

8
Fixing and Stopping   84

9
Faults and Repairs   95

Index   109

PREFACE

This pocket guide answers many questions on materials, methods and faults in the craft of plastering. It describes a variety of techniques, from basic solid plastering to the production of complex mouldings, and covers restoration and repair as well as new work.

It is hoped that the book will appeal to a wide range of readers seeking a handy and concise reference source. Although limitations on space make it impossible to deal with some of the more advanced problems, many of these can be solved by applying the basic principles described.

Plastering and general building students will find here the answers to many questions in the City and Guilds Craft Examination Paper, and also to numerous practical problems that will arise on site. Builders should find the book valuable both for the site office and for their professional examinations, particularly the sections on materials and faults. Also, do-it-yourself enthusiasts will be able to extend their practical skills, and learn something of the scope and intricacy of this demanding but satisfying craft.

1

MATERIALS

What is plaster?

Although lime and cement mixes, through common usage, are
often referred to as plaster or plasterwork, the material *plaster*
is manufactured from the mineral *gypsum*. This comprises
calcium sulphate and water chemically combined in a crystalline
form. By varying the method of manufacture and by the
addition of accelerators and retarders, four main classes are
produced: hemihydrate plasters, classes A and B, and anhydrous
plasters, classes C and D.

How do gypsum plasters set?

They set by chemically combining with water to grow gypsum
crystals. These interlock with each other, with the surface of
any aggregate present and with the surface to which they are
applied. During this change of state the plaster expands and
heat is given off by the chemical reaction.

What types of plaster are there and how are they classified?

1. *Plaster of Paris.* This is a hemihydrate, known as casting
plaster, Class A. It is made by crushing high-quality gypsum to
a powder and heating it until three-quarters of the water is
driven off. Unadulterated, it sets in approximately 20 minutes
and is used neat for running mouldings on the bench and for
fixing and stopping fibrous plasterwork. Mixed with lime, it is
used for small patches of setting and for running and stopping
in situ mouldings. With a retarder added by the plasterer, it is

used neat for casting in fibrous plasterwork and mixed with sand for small patches of undercoat in solid plastering. Coarse, fine and superfine grades are available, as well as special hard casting plasters.

2. *Retarded hemihydrate plasters, class B*. This plaster is produced from a slightly less pure gypsum and has a retarder added by the manufacturer to control the setting time. All undercoat plasters belong to this class and are known as type 'a'. The finishing plasters are known as type 'b'. Undercoat plasters for mixing with sand and finishing plasters to be used neat or with the addition of up to 10 per cent lime comply with BS 1191 Part 1: Gypsum building plasters. Lightweight undercoat and finishing plasters comply with BS 1191 Part 2: Gypsum building plasters.

3. *Anhydrous plasters, class C*. These are available as finishing plasters only. Probably the best known is Sirapite. It is made from less pure gypsums, which can contain up to 25 per cent impurities. It is crushed to a powder and heated to drive off approximately 90 per cent of its water content. Its set is slow and continuous, and it should only be used on well-keyed, moderately hard undercoats.

4. *Anhydrous plasters, class D*. These are available as finishing plasters only. Probably the best known is Keene's cement. It is made from high-quality gypsum of at least 90 per cent calcium sulphate. It is crushed and then heated over a longer period than is class C so that as much water as possible is driven off — somewhere in the region of 95 per cent. The resulting plaster needs an accelerator, which is added by the manufacturer, to start the setting process. This all combines to produce a plaster of high quality that is harder, takes longer to set and should only be applied to strong, well-keyed backgrounds.

What is ordinary Portland cement (OPC)?

It is a material that sets hard when in contact with water. It is manufactured by heating calcium (usually in the form of lime-stone or chalk) with silica, alumina and iron oxide (usually in the form of clay or shale) until they fuse together at a temperature of around $1400°C$. The resulting clinker is then ground to a powder with $4-7$ per cent raw gypsum to retard the early set.

How does OPC set?

OPC sets by reaction with water. When it is mixed with water its molecules combine with those of the water and form crystals. These crystals interlock with each other and with the surfaces of any aggregate added to the mix. The material possesses an early set, known as the initial set, which can be made use of when OPC is mixed and used quickly in small quantities, and a final setting time of not more than ten hours. Further hardening is brought about by continued reaction, and is rapid in the early stages but slower with the passage of time.

Depending on what it is to be used for, the cement mix will need to be cured to sufficient strength. Curing is effected by keeping the mix damp to allow the hydration of the cement to continue.

When is OPC used in plastering?

OPC is plastering's hardest-setting material and is used in:
(a) floorlaying, toppings to stairs, roof screeds, etc., with sand and granite for strong mixes and with vermiculite for light-weight screeds;
(b) in all external mixes, mixed with sand for undercoats, finishes and mouldings;
(c) in internal undercoats, mixed with sand to receive the harder finishing plasters, such as Sirapite and Keene's cement;
(d) for all waterproof plastering, with sand and waterproofing agent;
(e) for casts to be used externally, with sand or crushed stone;
(f) for grouts, neat, mixed with water.

What is masonry cement?

It is an OPC with a plasticiser included. It produces a highly workable mix without the addition of lime.

What is hydrophobic cement?

It is a Portland cement produced to stay fresher longer during storage in damp conditions. It has a waterproofing agent added to it during grinding. This coats every particle and makes it impossible for the cement to be mixed with water. The coating is removed during the normal process of mixing with sand, and the cement can then be wetted. The only effect the water-proofer has on the mix is to increase its workability slightly.

What is water-repellent cement?

It is a proprietary brand containing a waterproofer. Water-proofing compounds may be added to OPC mixes.

What is rapid-hardening cement?

Rapid-hardening Portland cement is ground more finely and hardens more rapidly after setting. *Extra-rapid-hardening* Portland cement is rapid-hardening cement with an accelerator added during grinding. This speeds up both setting and hardening.

What are coloured cements?

There are two categories of coloured cement: normal cement/sand mixes must be disguised with dark, strong pigments; brighter, pastel mixes can be produced by employing white Portland cement and silver sand. White cement is OPC with white china clay used in its manufacture.

What is builder's lime?

It is pure lime that is suitable for plastering and is known under the names of white lime, mountain lime, chalk lime and high-calcium lime. It is sold in the form of calcium hydroxide (slaked or hydrated lime), an off-white powder.

Calcium carbonate (limestone or chalk) is heated to drive off the carbon dioxide and so form calcium oxide (quicklime). This is slaked to hydrated lime by the addition of water. As the water combines chemically with the quicklime, the latter breaks down into a fine, dry powder.

These limes will not set in contact with water and so are known as *non-hydraulic* limes. For the plasterer, they are prepared and stored wet, in the form of lime putty or 'coarse stuff' (sand-lime mortar: see below).

Limes produced from limestone or chalk containing clay will set in contact with water and are known as *hydraulic* or *semi-hydraulic,* depending upon the extent to which this takes place. Magnesian or dolomitic limes are made from a limestone containing magnesium carbonate. Most are hydraulic; a few are semi-hydraulic.

How does plasterer's lime harden?

As explained above, plasterer's lime is non-hydraulic and supplied as a wet mix, ready for application. It hardens when the water added during slaking is given up, and carbon dioxide from the air recombines with the material so that it reverts to its original form. (Hydraulic limes have a Portland-cement-like material in them that is produced by their clay content combining with some of the calcium. They undergo an initial set when this material sets. Further setting is as for non-hydraulic limes.)

What is lime putty?

It is a material of putty consistency, made by adding hydrated non-hydraulic lime to water, stirring and leaving to soak.

What is coarse stuff?

Coarse stuff is a mixture of lime and sand that can be bought premixed to proportions specified, ready for the addition of cement or gypsum plaster.

5

When is lime used in plastering?

1. In sand and cement mixes as a plasticiser. In a volume of normal plastering sand, one-third consists of voids which must be filled to produce a workable mix. Therefore, when mixes weaker than one part cement to three parts sand are required, the difference is made up with lime (e.g. 1 part cement : 1 part lime : 6 parts sand, or 1 part cement : 2 parts lime : 9 parts sand). In all but severe conditions, these weaker, more porous mixes are more durable and weather-resistant.
2. To produce a fatty, sticky, butter coat on to which dashing mixes are thrown.
3. Mixed with non-lightweight class B and C gypsum finishing plasters to produce more workability and higher porosity on waterproofing undercoats.
4. Mixed with plaster of Paris for running mouldings *in situ*: (a) for the running screeds; (b) as a filler to thicken the plaster to a workable, putty-like consistency when running; (c) to reduce the expansion of the plaster. This lengthens the running time and prevents the moulding from splitting away due to expansion.
5. In repair and restoration when the original mixes and materials are required.

What sand is used for plastering?

Sand for plastering should be sharp, angular and virtually free from all impurities. This is the sand supplied as plastering sand from builders' merchants. In contrast, bricklaying sand is 'soft', i.e. individual grains are round. This tends to produce a structurally weaker set mix, which may shrink and crack when spread in sheet form.

What are the uses of sand in plastering?

As a filler in all cement and cement/lime plastering externally (i.e. undercoats, finishes, *in situ* moulded work, castings); internally in cement and cement/lime undercoats to receive

6

gypsum plaster finishes; both externally and internally with cement for floors, pavings, staircases, etc.

What is gauging?

In plastering, what was almost certainly the term describing the *proportioning* of materials in a mix has now come to mean the process of *mixing*. To 'gauge' or 'gauge up' is, therefore, to mix.

What is bulking?

Sand will only compact to its smallest volume when either bone-dry or saturated. When it is merely damp it is prevented from compaction by the grains' sticking together. A maximum bulking effect can account for one third of the volume. Such a mix would shrink by one quarter on the addition of water, reducing a $1 : 1 : 6$ mix to one of $1 : 1 : 4\frac{1}{2}$. Bulking can be detected by taking a level bucket of sand and stirring it slowly into a similar bucket filled with water. If the saturated sand does not reach the top of the bucket, the deficiency can be allowed for in proportioning.

How are sanded mixes gauged?

It is essential that the volumes of materials in each batch be accurately proportioned. In small jobs a bucket can be used for this purpose, but larger jobs call for the construction of a wooden 'gauge box', which consists simply of four sides. Material is placed in the box, struck off level with the top, and so produces one volume of the mix.

When all the volumes of the materials for the mix have been measured they are blended together dry. For thorough mixing by hand the mix should be turned over three times dry and three times wet. 'Turning over' takes the form of moving all the material to one side with a shovel, moving it back and then moving it to one side again. At this stage the mix should be of uniform colour with no lumps or pockets of aggregate or binder. A hollow like a volcano crater is made in the top of the pile and the water added. The dry material is then mixed with the water, starting with the inside of the ring. When enough water has been incorporated to make the mix of the right workable consistency the process of 'turning over' is again performed.

How are gypsum plasters gauged?

Plaster of Paris and all finishing plasters are sprinkled into clean water and allowed to soak. They are then stirred and may have a little more plaster added to bring them to a workable consistency. The premixed lightweight undercoat plasters, containing the lightweight aggregates perlite and vermiculite, are added to clean water in a bucket or mixing tray and mixed to a usable consistency — something like thick porridge. Plaster and sand mixes are gauged as described in the previous answer.

How should sacks of plaster, lime and cement be stored for use?

To keep the sacks dry, they should be stored on a timber platform, away from walls, and should be covered to protect them from the rain. All materials should be used in rotation to avoid excessive deterioration due to prolonged storage. The stacks should also be protected by barriers or situated where the sacks will not be torn.

How should plasterboards be stored?

The boards should be laid flat, one on top of the other, to form a stack not exceeding 0.9 m in height. They must have a level, dry base: timber across the width, spaced at 400 mm centres, or a sheet of polythene on a flat floor. They must, of course, be protected from any damp.

What plastering materials can be applied to which backgrounds?

Gypsum plasters can only be used internally because they must be protected from the weather. The factors that govern the suitability of a mix are adhesion, strength and corrosiveness.

Three types of lightweight premixed plaster are available; these should all be finished with premixed lightweight finish.

1. Browning, class B type a1, for application to all the common backgrounds that have natural mechanical keys (brickwork, blockwork, etc.).

2. Metal lathing, class B type a2, for application to backgrounds subject to corrosion (metal lathing materials, Newtonite waterproofing lath and woodwool slabs).

3. Bonding, class B type a3, for use when extra adhesion is required due to a lack of mechanical key (plasterboards, surfaces treated with PVA adhesive, smooth concrete, etc.).

A board-finish grade of plaster, class B type b2, is available for one-coat plastering on plasterboard and fair-faced concrete.

Cement-bound mixes produce the strongest material and should not be applied to backgrounds weaker than themselves. Such a situation would result in an 'icing on a sponge cake' effect if the background were to flex. A 1 cement : 1 lime : 6 sand mix can be applied to all solid backgrounds, but should be preceded on smooth concrete by a spatterdash key coat. The most suitable finishing plasters are Sirapite or Thistle finish. A 1 cement : 3 sand mix should be applied only to the strongest backgrounds when the finish is to be Keene's cement.

In the case of fire, what will be the effect of extreme heat on internal plastering?

The main effect of extreme heat on applied plastering will be the breakdown of adhesion between the background and the backing coat of plaster. This is thought to be greater when the backing coat contains sand; the lightweight aggregates vermiculite and perlite provide greater adhesion.

However, under BS 476 all gypsum plasters and plasterboards are defined as non-combustible. From this it is obvious that all these materials will reduce the speed of spread of fire both within a building and from building to building. This resistance is due mainly to the proportion of chemically combined water within the gypsum plaster. This will have to be driven off before the breakdown occurs.

2

SOLID PLASTERING

What is solid plastering?

It is the practical application of wet plaster mixes to a background, and is also referred to as work carried out *in situ*. It refers to all types of wet plastering: internal, external, floor screeds and pavings, and all moulded work run in position.

How is solid plastering applied?

There are two methods used for the application of wet plasters. First, there is a variety of spray machines which in general mix, pump and spray cement, lime, sand or gypsum plaster on to a prepared background.

The second method is the more usual one of hand application with the assistance of a hand hawk and laying-on trowel (Fig. 1). When using these tools, a right-handed person will

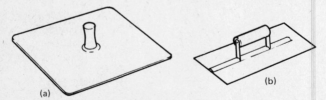

Fig. 1. Basic tools: (a) hand hawk, (b) laying-on trowel

hold the hawk in his left hand and the trowel in his right. The index finger of the right hand should be pressed against the front shank of the trowel, with the back of the hand upper-

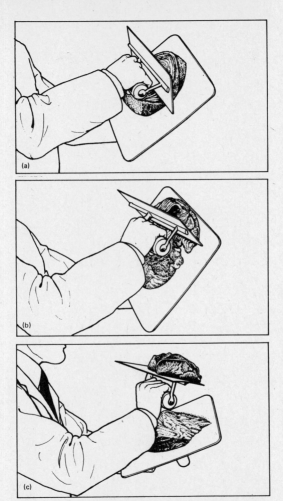

Fig. 2. Transferring plaster from hawk to trowel

11

most and the toe of the trowel pointing left. The index finger and thumb of the left hand should be in contact with the underside of the hawk and the handle just loosely held, not tightly gripped.

The material must be mixed to the right consistency, and placed close to the working area on a spotboard and stand. It can be removed in workable loads from the board to the hand hawk by placing a corner of the hawk under the board and pushing with the trowel sufficient material on to the visible three-quarters of the hawk face. From here it is carried to the working area and transferred from hawk to trowel (Fig. 2) and from trowel to background (Fig. 3). It is not necessary or

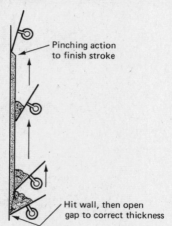

Pinching action
to finish stroke

Hit wall, then open
gap to correct thickness

Fig. 3. Application of solid plastering

desirable to use the hawk load up in one. A plasterer will probably carry two, three or even four trowel loads in one hawk load.

The application of backing coat and setting coat will vary according to need, since the latter will be of a thinner consistency.

12

What is a background?

It is any surface to which the first coat of plaster is applied. This coat may vary from a render or scratch coat in three-coat work, the floating coat in two-coat work, or the finishing coat in one-coat plastering.

For all practical purposes, backgrounds are classified under two headings: *high-suction* and *low-suction*. Within these two categories there is another, *lathings*. Most of the latter are low-suction backgrounds. However, they require fixing instructions and in some instances specific practical operations when being plastered.

It is essential to be able to identify the various types of background. The gauge or mix of the backing or first coat will be controlled by this, as will the type of premixed plaster used or specified.

What is a low-suction background?

It is a background that absorbs little or no water from the mix. Engineering bricks, concrete, bonding agents (polyvinylacetate, PVA), cork and the majority of lathings (e.g. plasterboard, plaster lath, metal lath, K lath and Newtonite lathing) are all low-suction backgrounds.

What is a high-suction background?

This is a background that will absorb water from the plaster mix as it is being applied. Common clay bricks, lightweight building blocks (Thermalite, Duroc, etc.), clay tiles or pots are examples of this type of background.

How should a background be prepared before plastering?

Brickwork of all types, blockwork and the like must be dry brushed after all joints have been raked out. When free of all dust and dirt, the entire area should be dampened prior to being plastered.

13

Concrete must be clean, free from dust, dirt or mould oil, and keyed. The keying may be formed by hacking, spatter-dashing, or applying a patent liquid adhesive in accordance with the manufacturer's instructions. Once again, for perfect adhesion concrete surfaces should be dry brushed then lightly dampened before plastering begins.

All lathings must be correctly fixed and checked before plastering commences.

What is spatterdash?

It is ordinary Portland cement and sharp clean sand, mixed to a thick slurry in the ratio of 1 : 2. Spatterdash has two basic functions in solid plastering. First, when thrown on to strong dense backgrounds it provides a first-class key for the backing coat, internally or externally. Secondly it provides an even suction. The throwing action makes for better adhesion and provides a deeper key.

What is a backing coat?

A backing coat may be one of two things, *render coat* or *floating coat*. It is, however, always the first coat of any type of undercoat applied to a background. Under the first heading, render coat, it consists of one rough coat of undercoat plaster. The average thickness should be 8 mm, and it is not ruled or straightened in any way. As soon as it is firm enough it is keyed horizontally with wavy lines by a comb or wire scratcher.

Under the second heading, floating coat, it is applied either directly to the background (two-coat plastering) or on top of a hardened render coat (three-coat plastering). In the first case it should average 10−12 mm thickness, while in the second it should be 8 mm again. It is ruled flat and straight, and keyed by devil float or by scratcher to receive setting or finishing.

The main functions of a render coat are to help make backgrounds such as metal lathing rigid, and to present a well-keyed backing for a floating coat. The main functions of a floating coat are to present a good, true, flat surface for setting internally and finishing externally, and also to provide a uniform suction for working upon.

When is a render coat required and how is it applied?

Render coats, scratch coats and 'pricking up' are always essential when any form of solid plaster is being applied to a metal lath background. Three-coat plastering (render, float and set) is also used for special purposes, e.g. internal water-proofing. It is frequently used as an additional backing coat in external platering and for external waterproof renderings.

The render coat is applied on average to a thickness of 7–10 mm. It is applied by trowel to a reasonably even surface and keyed by comb scratcher.

Why is it necessary to key backing coats differently?

It is always necessary to key the render coat. The reason for this is that the keying will have to support an 8 mm thick floating coat plus whatever finish is then applied to the floating coat.

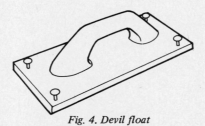

Fig. 4. Devil float

The keying of the floating coat will depend upon whether the finishing is a light, thin-coat material or a thick, heavy material. *Internally* the finishing or setting is nearly always light and thin, so the keying will be carried out by a devil float (Fig. 4). This is laid flat on the floated surface and rubbed over

15

the surface in a circular manner. The nail points that project just below the underside of the float will form indentations or keyings.

Externally the finish may well be heavy, so a comb or wire scratcher is used (Fig. 5). However, no keying is required in the

Fig. 5. Wire scratcher

case of light finishes (i.e. the resin and plastic materials, also Tyrolean finish).

To sum up, backing must be keyed differently mainly because of the various weights and thicknesses of a wide range of materials.

What are dots and screeds?

'Dotting and screeding' is a method used to obtain a plumb, true and level surface in a floating coat. See Fig. 6.

Dots are usually mounds of the floating-coat material applied to the background; usually a 75 mm long batten (or similar) is embedded in the face of each dot. A plumb rule, level, square or gauge is used to ascertain the final position and accuracy of dots when they are being plumbed down, squared in or levelled. Alternatively, dots may be nails driven into the background and used as grounds for the same operations.

Screeds are strips of various types of plaster mix. They are laid between dots or applied on their own, to act as grounds for ruling off the plaster applied between them.

How is a floating coat applied?

The application of a floating coat to a wall or ceiling is basically the same as for all backing coats. However, the purpose of a

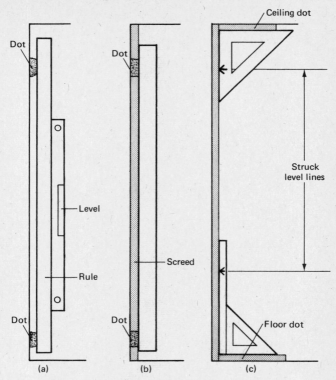

Fig. 6. Dots and screeds: (a) plumbing down a pair of dots using parallel rule and spirit level; (b) floating vertical screed from the dots; (c) squaring-in ceiling and floor dots from a level line, using marked squares

floating coat is to provide a flat true surface for the finishing. The following method is usually adopted for work of good standard. With dots and screeds or grounds already in position, the floating coat is laid on to these and then ruled flat with

17

either a floating rule or metal featheredge (Fig. 7). The grounds may be wooden at skirting and ceiling line, and metal angle bead or pre-fixed wooden rules at external angles. Most materials are given specific thicknesses within the specification, and all should be applied in several thin layers with the same mix, rather than out to full thickness at one go. After the

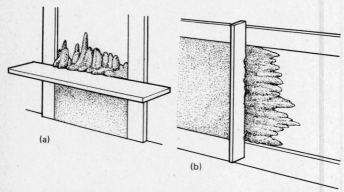

Fig. 7. Ruling floating coat: (a) from vertical screeds, (b) from horizontal screeds

rule has passed over the entire surface, wooden dots should be removed, extra material added to any hollows, and the rule used once again. When a satisfactory surface has been obtained, flush to all screeds or grounds, it should be allowed to 'pick up' and finally keyed to receive the finish.

What is a finishing coat?

The answer to this question must be in two parts, distinguishing basically between *internal* plastering and *external* finishes or renderings, though it is possible to apply the latter internally.

For internal work, the finishing or *setting* coat will be a thin coat of gypsum plaster, applied to a keyed backing coat. It is also possible to use a specially prepared thin-coat plaster,

18

which can be applied direct to a suitable background. The main function of a setting coat is to supply a satisfactory surface for decoration.

External finishes vary from thin-coat plaster finishes to heavy textured work such as English cottage, fan texture and the various dashings both wet and dry. The textured finishes are many and varied, while the dashings are restricted to the two types mentioned; see page 31. However, different aggregates may be used with dashings to obtain different appearances.

How is the setting coat applied?

The application of this coat will depend upon (a) the material being used for the setting coat, and (b) the method considered best for the job in hand.

To deal with (a) first, an internal plaster finishing or setting coat will be carried out in one of the following six materials.
1. Class A, B or C gypsum plaster, with up to 25 per cent lime putty added.
2. Class B board or wall finish, neat.
3. Class C gypsum plaster, neat.
4. Class D gypsum plaster, neat.
5. Premixed lightweight gypsum plaster, neat.
6. Thin-coat gypsum plaster.
In all cases, specifications and manufacturer's instructions must be adhered to.

With reference to (b), the method chosen will depend upon various technical factors. However, a setting coat is always applied using one of the following methods:
1. Three thin coats (trowel, float, trowel).
2. Two thin coats (float, trowel).
3. Two thin coats (trowel, trowel).
Of the materials listed above, numbers 1, 2, 3, 4 should be applied to an average thickness of 3 mm, while 5 and 6 should always be 2 mm or less. In practice the trowel, float, trowel method produces the flattest surface, and is always used when applying class D or class C plaster neat.

The method of application is to lay a tight coat of plaster over the entire area of backing. This is ruled in flat with a featheredge rule; then, all hollows having been filled in, it is ruled off again. The second coat is applied with a skimming float (Fig. 8). It must be applied in long, full, vertical sweeps

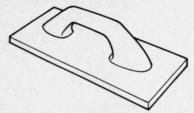

Fig. 8. Skimming float

over-full on the way up, with the excess removed on the way down. This ensures a perfectly flat surface. A very tight third coat is next applied by trowel and laid on from left to right. This coat must result in a flat smooth finish. As the plaster sets it is brought to a hard smooth surface by being trowelled, lubricated with a little water until a satisfactory finish is obtained.

The alternative methods (2 and 3) either eliminate the first trowel coat and commence with the float, or eliminate the second coat (float) and consist of two trowel coats. These are smoothed in the same way as in method 1.

How are internal angles formed?

There are several methods used when forming internal angles in gypsum plaster. Probably the simplest is to allow one wall to set hard before applying the setting coat to the adjacent wall. All that is then required is for the toe of the trowel to be held flat on to the wall surface and passed down the entire length of the angle.

Another method is to lay the setting coat on to both walls

20

and rule out vertically both ways from the angle. When the plaster is hardening, lightly pass a crossgrain float (Fig. 9) up and down both angle surfaces; a smooth surface is obtained with an internal-angle trowel (Fig. 10).

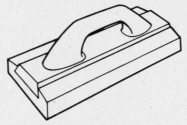

Fig. 9. Crossgrain float

Fig. 10. Internal-angle trowel
(twitcher)

Whichever method is used, care must always be taken to see that the backing coat has been cut out clearly at all internal angles. Lay the trowel flat on the floated surface and allow one tip of the toe to cut into the adjacent surface at the angle. Repeat the process on the other wall and this will clean the angle out perfectly.

How can external plaster angles be protected from damage?

Plain angles, pencil round, were protected traditionally by the use of hard plaster angles (Fig. 11). In this method, a

50 mm strip of cement/sand would be applied on either side of the arris; when set this would be finished in Keene's cement (class D plaster) or Sirapite (class C plaster). This usually took place in public buildings because the lime plasters of the day were not sufficiently hard to withstand knocks.

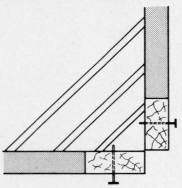

Fig. 11. External angle: battens used as floating screeds, then removed and the hard plaster angle floated

Today, the metal angle bead (Fig. 12) has eliminated the need for hard plaster angles. The normal metal angle bead consists of a hollow metal bead with two metal lath wings. The plasterboard or one-coat-finish metal bead has two strips of perforated metal on either side of the bead in place of the metal lathing.

How are square external angles formed?

In internal plastering they are usually formed with the aid of metal angle beads. These are bedded at external angles in either class B gypsum plaster or premixed lightweight bonding plaster. One method is to apply 30 mm plaster dabs at 600 mm intervals; another method is to apply the plaster the full length of the

22

angle. The bead should be pushed straight with a long rule and plumbed, squared and margined. When the bedding plaster has set, the walls, piers, beams, ceilings, etc. may be floated, using

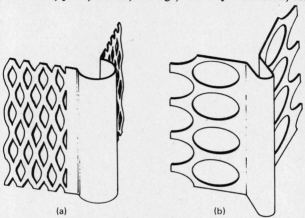

Fig. 12. Metal angle bead: (a) normal, (b) plasterboard or one-coat finish)

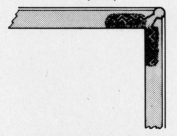

Fig. 13. Use of metal angle beads
to form external angle

the beads as ruling grounds. Once the floating has become sufficiently firm, it should be ruled hard to a level just below that of the bead nosing; the surface will be brought up to the nosing level by the setting coat (Fig. 13).

The normal bead allows the correct thickness for two coats, floating and setting. The plasterboard grade is meant for one-coat finish to plasterboard and is nailed in position.

Externally and internally, where angle beads are not used the wooden-rule method is general. The rules are fixed to the reveals so that the main surface may be floated, using the rules as grounds; see Fig. 14(a). Due allowance should be made for the correct thickness of plaster. Once hard, the rules are removed and fixed in the same manner to the floated surface; see Fig. 14(b). The second side is then floated, and when complete and firm the rule is removed once again. A floating rule is then passed gently over the angle to remove any selvedge. Both internally and externally, the finishing coat to all external

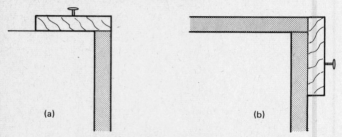

(a)   (b)

Fig. 14. Use of wooden rule to form external angle

angles backed in this way may be applied in precisely the same manner, switching rules; alternatively, it may be done without the aid of rules altogether, or in certain cases with hand-held temporary rules.

What is a reveal?

These are usually found at openings, both door and window, where they are the narrow wall surfaces, generally at right angles to the main walling. They may be vertical or horizontal (in the latter case often being referred to as a 'head'), and are also found at the sides of a chimney breast and at the sides of plain piers.

What is the best way of ensuring that all reveals are square and marginable?

Using either of the methods described for external angles, i.e. metal bead or wooden rule, a *reveal gauge* is necessary to obtain a true margin around the frame, door or window. A gauge also is required to ensure that the reveal in itself is perfectly marginable. Fig. 15(a) shows a single appliance that will do both jobs.

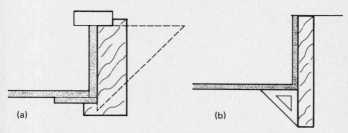

(a)   (b)

Fig. 15. Squaring-in a reveal: (a) from back surface, (b) from forward surface

A square must be used to check the accuracy of the gauge. Also, if the back section is not a door or window frame, the square may be used to get a square reveal from the back wall. The method used to check the square from the facing wall is shown in Fig. 15(b).

What is meant by the term acoustic plastering?

Acoustics means the science of sound, and acoustic plastering means the control of sound by the application of a specific plaster. In the past, several types of site-mixed plasters were used to achieve this end, but nowadays wet acoustic plastering is carried out using a standard ready-mixed acoustic plaster. This consists of a retarded hemihydrate plaster with graded pumice added to provide a porous aggregate.

Before applying the acoustic plaster, all backgrounds must be prepared and floated in the appropriate grade of lightweight

25

aggregate gypsum plaster to a thickness of 10–12 mm. This backing coat must be ruled flat and cross-keyed with a comb scratcher. It should then be allowed to dry out sufficiently to provide suction for the application of the acoustic plaster.

This must be applied neat and in two applications, each coat being 6 mm thick. The first of these coats must be ruled flat, deeply scratched and allowed to dry, once again to provide sufficient working suction (moderate suction is essential for both coats of the acoustic plaster). The second coat is then applied and flattened with a wooden float. It must be left in this open-textured condition and at no time finished by steel trowel. To assist in the production of the desired open texture, some craftsmen cover the float face with either carpet or cork. The entire area must be finished in one operation and at no time should acoustic plaster be applied below a two-metre dado height. Being a relatively soft and open-textured material it is easily damaged and picked.

This kind of plastering is normally carried out in the larger type of public building. To improve sound insulation in smaller domestic buildings the following recommendations are occasionally made. Where cavities exist in plasterboard partitions and dry linings, and between ceiling and floor, they should be filled with glass fibre or mineral wool. All surfaces should be plastered using two coats of lightweight premixed plaster, the appropriate grade for the backing being finished in neat-finish plaster. Also, acoustic tiles are recommended for workrooms and, at times, kitchens. These may be fixed to a solid background with a recommended adhesive.

What is a gig stick?

It is an appliance used by a plasterer to form curved work, plain or moulded, in much the same way as the leg of a compass. It may be attached either to a running mould and used to run curved mouldings (see page 59), or to a square or batten to assist with the formation of run screeds (see below). At the pivot end it may have a fish-tail shoe; these are usually considered better for solid work (Fig. 16). Alternatively it may

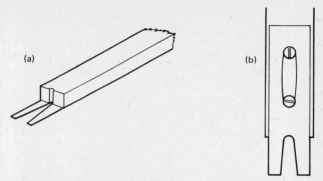

Fig. 16. Gig stick with fish-tail shoe; (b) shows adjusting screws

have a side shoe; this method is normally used in bench work (Fig. 34).

Gig sticks are also referred to at times as *radius rods;* these are used for setting out.

How are curved surfaces plastered accurately in solid?

This is done by one of two methods: (a) run screed, or (b) pressed screed.

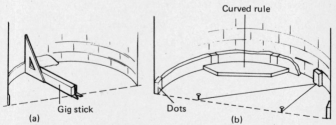

Fig. 17. Curved surface: (a) run screed, (b) pressed screed

When using method (a), screeds of the backing material are run in position by a square or batten attached to a gig stick; see Fig. 17(a). This pivots from the centre/centres, to produce

27

perfectly formed screeds. Where it is impossible to use a gig stick, i.e. where the curve is not circular, a batten and focal points are used.

With method (b), one must still be able to locate centres and/ or focal points. From these, dots are positioned using a long line or a batten stretching from the points to the dots; see Fig. 17(b). Once the dots are positioned correctly, screeds are pressed in place between the dots as in normal floating; the difference is that a purpose-made template, cut or cast to the curve, is used to press the screeds into the correct shape.

In both instances, once a perfectly curved backing coat has been formed the finishing should cause few problems. With correct application this should automatically follow the lines of the backing. However, there are times when it may be necessary to form screeds for the finishing. If so, the procedure is the same as for the backing. At other times, and where the area is small, under a bay window for example, it may be possible to run the entire surface out using method (a).

How are plain arches formed in solid plaster?

These may be formed in one of two ways: (a) with templates, or (b) run with a gig stick.

In (a), one requires templates cut to the arch outline. These may be hardboard, plywood or fibrous plaster. The main wall surface should be floated, the template/templates fixed in position, and the soffit floated with a rule. The rule cuts the backing coat back to allow the finishing to fall in line with the templates; see Fig. 18(a).

For (b), the gig stick is a stiff piece of batten with the arch arris cut out at one end. The radius is measured back from the return angle, and the stick is cut accurately so that a fish-tail may be attached to the other end. See Fig. 18(b). A stretcher with a pivot is fixed across the opening, and the gig stick runs from this. First, a floating coat is applied to the main walling and soffit. To assist with this an extra small batten (floating guide) may be attached to the gig stick so that it will reach across the soffit. A small finishing member (i.e. a strip of

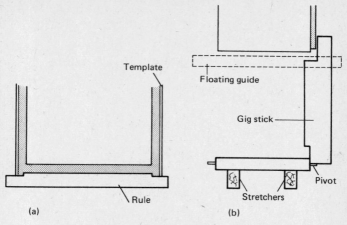

Fig. 18. Plain arch: (a) templates, (b) gig stick

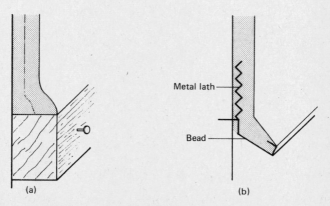

Fig. 19. Bell cast: (a) removable batten, (b) external render stop

29

finishing coat) is then run at the arris, to act as a finishing screed for both wall face and soffit.

What is a bell cast?

This is the lip formed at the bottom of external renderings or over doors, windows, etc., to prevent water from running down the remaining surface underneath. It forms a drip member. It may be formed either by positioning a thick rule to work to *or* by fixing an external render stop and working to this (Fig. 19). The latter is a purpose-made metal bead made from galvanised sheet steel. It may be either nail-fixed or bedded with dabs of cement and sand.

How is a cement plainface finish obtained?

A cement plainface may be a finish, or it may be a backing. In the first instance it is more usually termed 'float finish'; in the second it is a perfect backing for Tyrolean machine-applied finish, and also for the natural aggregate/resin finishes.

Fig. 20. Jointer

Apply as in floating, rule level and flat, then rub a large wood or plastic float over the entire surface. This will result in a flat, stone-like texture that can be left as it is or marked out to represent finished blockwork. In the latter case it is often called 'ashlar', after the stonework that it resembles. The markings may be done by a jointer (Fig. 20) or a large square nail.

How is Tyrolean machine finish applied?

It is applied to a cement plainface backing, by a Tyrolean machine. This is a hand-operated machine with steel springs attached to a spindle. The springs contact an adjustable bar when the spindle is turned; this produces a flick-on action. The finish material is mixed wet and tipped into the machine. The operator should stand 0.5 m from the wall and more or less at right angles to it. The aim of the first coat is to produce a light, uniform, honeycomb texture. When this is firm, apply the second coat from 0.5 m, but this time at 45°. Once again, gradually build up the material to produce a slightly heavier honeycomb texture. When it is firm, apply the third coat, again from 0.5 m and either from an angle or straight on. Coats and joints must be staggered, and there should be no bald patches or heavy runs. Windows and doors must be well masked out.

How are the dashed finishes applied?

Both types of dashed finish need a flat, well-keyed backing coat, with just one difference. *Wet dashing* (also known as roughcast or harling) requires a backing coat that will give a reasonable controlled suction. *Dry dashing* (also known as pebbledash) needs no suction at all; in fact it is usual to add a percentage of waterproofing compound to a 1 : 3 or 1 : 4 OPC/ sand backing mix. The backing should be 1 : $\frac{1}{2}$: 4 or 1 : 1 : 6 OPC/lime/sand for roughcast.

To obtain a roughcast finish, a butter or thin coat of 1 : 1 : 6 is applied; then the roughcast (also 1 : 1 : 6, the six being three parts sand, three parts shingle) is dashed on to the butter coat. It is thrown on with a harling trowel (Fig. 21); the wrist is used to get a full spray and not just a blob of dashing on the wall.

For pebbledash, clean and dry pebbles are dashed on to a 1 : $\frac{1}{2}$: 4 or 1 : 1 : 6 butter coat, as follows. Buckets or boxes (Fig. 22) full of clean pebbles are placed adjacent to the area. A butter coat is applied to an area of approx 2 m². One plasterer should sheet out to catch the falling pebbles, and then with the dashing trowel commence to throw the pebbles on to

31

the butter coat. Again the wrist must be used to obtain an even fan-spread of pebbles. All dropped pebbles should be collected, washed and dashed again. Once a start has been

Fig. 21. Harling trowel

made, one man can keep just ahead of his partner, spreading, while his colleague dashes. A wooden float may be used to press the dashings lightly in.

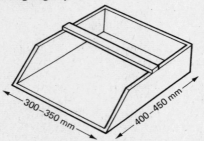

Fig. 22. Dry-dash box

In both types of work, external angles may be dashed free-hand or up to a hand-held rule.

How are natural coloured aggregate/resin finishes applied?

There should be a clean, flat, lightly dampened cement plainface backing with no keying. The trowel-finish material is usually applied by trowel as thinly as possible. The backing must not be visible at any point after the application of the finish. Some manufacturers supply a brush-applied primer.

32

The finishes are supplied in airtight containers, which should be emptied completely on to a perfectly clean spotboard. Here they can be thoroughly mixed and any unwanted material returned to the tin. All tools must be kept perfectly clean during application. Apply in strong short upward movements of the trowel, keeping the leading edge moving so that no joints are left. Leave no trowel marks, but should the material need further trowelling do this with a damp trowel and once only.

Do not mix water with the material. Do not splash water on applied material. Do not apply in frost or where it can be affected by rain.

How are external textured finishes obtained?

They are applied to a well keyed backing. The mix can vary from $1 : \frac{1}{2} : 4$ to $1 : 1 : 8$ (OPC/lime/sand), and the textures may be anything from the conventional fan through to almost any desired effect.

The mix is applied to a general thickness of 10 mm, and the texture is worked on to this coat while it is still wet. For a fan texture, press one corner of the heel of the float or trowel into the wet material and move the tool in a rippling movement from left to right, holding the point still. The result should be a fan quadrant; these should be applied so that they overlap slightly. For English cottage, apply haphazard dabs of the finish material over the finish coat already applied and still wet. Plenty of daubs, overlapping and crossing will give an old-world finish. Stippling with sponge, twigs or stiff brooms may be used to get the required finish.

Do not apply a finish coat that is stronger than the undercoat. Use the backing coat to control and give an even suction.

How is an external waterproof finish applied?

The background must be perfectly clean, well raked and lightly dampened. To improve adhesion a waterproof spatterdash should be applied to the entire area. This must be allowed to harden before the backing coat is applied. When applying this, it must be free of joints, applied to an average thickness of 10 mm and well keyed to receive the finish. Once more, the finish should contain no joints and average 10 mm thick. Where possible, a three-day lapse between coats is advisable, during which time the backing must not be allowed to dry out. The actual finish may be any of the recognised external renderings. However, because of the lack of suction in the waterproof backing coat, certain of these are impracticable and the recommended finishes are cement plainface (float finish) and pebbledash.

The usual OPC/sand mixes are: spatterdash 1 : 2 plus waterproofing compound; backing 1 : 2 or 1 : 3 plus waterproofing compound. Finish to the finish specification. The waterproofing is added according to the manufacturer's instructions; of the different types, a liquid is recommended. Alternatively, water-resistant Portland cement may be used to the mixes quoted above.

How is an internal waterproof finish applied?

Once again, start by spatterdashing on to a perfectly clean damp background. Where possible, no joints should occur in the backing coat, and internal angles should be coved. Where joints have to occur they must be staggered. In this type of work two undercoats are recommended, render and float. The same rules apply to both ceiling and floor if these are included in the work schedule, except that the floor should receive one coat only.

The floor screed will probably be 1 : 2 OPC/sand plus the waterproofer. It must continue for 25–50 mm up the wall. Here it may be rebated so that the second coat of wall plastering will cover it. Alternatively, in the case of a paving, it may be coved into a cement skirting.

Where a gypsum-plaster finish is required, class B, C or D plasters are all suitable. However, 20 per cent of lime is frequently used; this will allow the finish to 'breathe'. Condensation will occur and can only be overcome by good ventilation.

34

3

PLASTERBOARD

How is a plasterboard ceiling fixed, scrimmed and set in one coat?

Plasterboards are fixed across the wooden joists by 30 x 2.6 mm galvanised nails at 150 mm centres. The boards must be cut at the centre of the joists so that the next board can be nailed to the same joist with a 5 mm gap between boards at all edges and ends. The usual method of cutting plasterboards is to mark the required length, cut or score through the paper on one side, turn the board over and snap it off at the cut. The paper may need freeing, but the overall result should be a neat, clean, straight cut.

Next comes the scrimming. For this operation a roll of 75–80 mm canvas scrim and a sharp knife are required. Cut scrim to cover all joints between boards and between wall/ceiling angles; do not allow any overlapping. Class B board-finish gypsum plaster is then mixed and laid so that it straddles all joints by approximately 40 mm each side, thickness 2 mm, and pushed well into the 5 mm gap. When all joints and wall/ceiling angles have been covered in this way, start with the longest piece of scrim. Hold this in your left hand and the trowel in your right. Open the scrim out, and place across the joint; then press it in with the trowel, traversing the entire length of the joint, and trowel in. Continue until all joints are scrimmed, including wall/ceiling. With the latter, it is best to put half the width on the ceiling first, and press the wall half home when the ceiling half is completed. Make certain no overlapping occurs, and lay board-finish over all plasterboards *between* scrim — not over it, just yet.

Now with fresh plaster apply a level, flat coat over the entire ceiling, and rule with a featheredge rule or darby. Lay another tight coat over the ceiling as the first coat picks up; this should result in a nice flat surface that can be trowelled smooth as it hardens. The trowel should be lubricated with clean water as this final trowelling takes place, and trowelling should continue until a hard smooth surface has been obtained.

Points to note are that, where possible, end joints to boards must be staggered; the trowelling of the scrim must take place before any hardening occurs; and no scrim should be left dry. A float may be used for an extra coat of finishing if thought necessary.

What is meant by 'dry lining' and how is it applied?

Dry lining is a method of providing a lining to the interior of a building using plasterboard as a flush flat finish, a V-jointed decorative finish, or a cover fillet finish. There are two accepted systems for this type of work, the *bonded* system and the *metal furring* system.

How is dry lining fixed by the bonded system?

The bonded system consists of fixing bitumen-impregnated pads to the background by dots of plaster. This should provide a flat, true surface to which the plasterboards can be fixed. The background is first marked out vertically at 450 mm centres, then checked for flatness; all high spots should be noted. Two pads are embedded in Class B board-finish plaster at each angle, one 230 mm down from the ceiling, one 100 mm up from the floor. These are then plumbed down, and intermediate pads are lined in vertically at 1070 mm centres. The process is continued over the area at 1800 mm centres; when these dots have set, the intermediate horizontal dots are then lined in from them. Where the high spots have occurred it is essential that the maximum thickness of bedding plaster is 3 mm. Provided this operation has been carried out correctly, the pads should lap both boards where joints occur.

36

Before the boards are fixed, dabs of board-finish plaster are applied vertically between the pads. As the boards are pushed back on to the set dots, the board-finish will then spread out over the back of the boards. This will ensure good adhesion between the background and the board. Two-headed nails driven into the board edge pads will provide a temporary fixing until the plaster has set (Fig. 23).

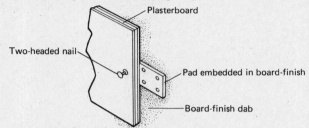

Fig. 23. Fixing dry lining using bonded system

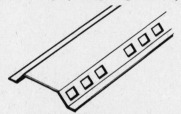

Fig. 24. Channel used in metal-furring
system of fixing dry lining

The two-headed nails are pulled out when the plaster has set, and the holes are filled with filler. Use pincers when pulling the nails out. This will ensure that no damage is done to the boards, and the nails may be used again.

How is dry lining fixed by the metal furring system?

The metal furring system consists of bedding 72 x 9.5 x 0.6 mm metal channel (cold-rolled mild steel, hot-dip galvanised; Fig.

37

24) to the wall vertically in drywall adhesive at 600 mm centres. The centres are lined in by a parallel metal straight edge to a continuous line struck on both floor and ceiling. Allow 9.5 mm from the highest point, plus the width of the straight edge, when striking the continuous line. Dabs of adhesive 200 mm long should be applied on both sides of the vertical lines about 25 mm apart and at 450 mm centres. The channel is then placed so that the two edges are well bedded in adhesive, and tapped back until the face lines up with the rest. Metal stops (furring channel in 150 mm lengths) are fixed horizontally between the channels at the top and the bottom.

Once the adhesive has set, the boards are fixed to the channel by 22 mm drywall screws at 300 mm centres. These have a Posidriv head and should be driven in just below the surface, but the paper should not be fractured.

How are the joints treated in dry lining?

The jointing for both systems will depend upon the type of board used. Square-edge boards are butted and covered, bevel-edge boards provide a V-joint decorative feature, and taper-edge boards are made flush and seamless by taping and jointing.

Manual jointing is carried out in three basic steps. First the joints are filled with Gyproc joint filler. A continuous band of jointing tape is next applied over the filler. Then a new layer of filler is applied over the tape. Any surplus must be cleaned off before any set has taken place. Once it has set, check and refill if necessary.

Gyproc joint finish is used to obtain a finish over the whole area.

How can the thermal insulation of a building be improved by the plasterer generally?

In several ways, including the use of thermal plasterboard, Gyproc insulating wall board and wet plastering.

Thermal plasterboard consists of Gyproc plasterboard

bonded to self-extinguishing expanded polystyrene. It is obtainable in the ivory grade surface for decorating and the grey grade for plastering, and in three thicknesses (22.2, 25.4 and 32.0 mm). It is ideally suited to ceilings where a high degree of thermal insulation is required.

Gyproc wallboard, insulating grade, also improves thermal insulation. Again, the wallboard is available in the ivory and the grey grade, and it has a veneer of polished aluminium foil on the reverse side to act as a reflector when facing inwards in a cavity. It may be used in ceilings, partitions, as dry lining or for encasing steel beams and columns. The exposed surface may be plastered or left for decorating. In general, the performance of all plasterboard backgrounds will improve with plastering, using the appropriate grade of undercoat plaster and neat lightweight premixed finish plaster.

To all solid backgrounds the application of two coats of plaster will again improve the thermal insulation values. The lightweight premixed plasters are usually recommended, because the aggregates perlite and vermiculite are considered to have greater thermal value than sand.

4

FLOOR SCREEDS AND PAVINGS

What is a floor and roof screed?

It is a mixture of ordinary Portland cement and sand applied
to a sub-floor and eventually covered with yet another material.
It is not a finished floor in itself, and the materials applied to
the upper surface of a floor screed may be tiles, clay, plastic,
lino or cork, wood blocks, terrazzo and many others.

A roof screed is similar: it can be exactly the same as a floor
screed, or it may have a lightweight screed laid between the
sub-floor and the topping. This supplies additional heat and
sound insulation to the building.

What is a paving?

This is a finished floor laid by the plasterer. A cement paving
consists of Portland cement and sand, while a granolithic
paving consists of Portland cement and granite. The latter is
generally made non-slip with the addition of carborundum
dust to the wet surface of a newly laid paving.

What are monolithic, bonded and unbonded floor screeds and pavings?

1. *Monolithic* or *one floor*. The screed or paving is laid directly
on to a green concrete slab floor within a few hours of the
concrete being laid. Minimum thickness is 20 mm.
2. *Bonded* or *separate floor*. This is laid to a hardened base
that must be clean, damp and grouted. Minimum thickness is
40 mm.

3. *Unbonded*. There is no adhesion between sub-floor and topping. In fact the topping is always laid upon either heavy PVC sheet or waterproof building paper. Minimum thickness is 75 mm.

What preparations are necessary to sub-floors prior to laying?

1. Monolithic: remove all laitance and divide the area up into bays of approximately 15 m².
2. Bonded: clean, saturate and grout the sub-floor. Bay sizes 15 m². (Grout = neat OPC slurry.)
3. Unbonded: cover sub-floor with PVC or building paper. Turn-ups at the perimeter 150 mm and laps at all points 100 mm. Bay size not more than 10 m². May be reinforced with galvanised chicken wire or similar.

How is a floor screed or paving applied?

1. Monolithic: to a clean, green concrete base.
2. Bonded: to a clean, wet, grouted sub-floor.
3. Unbonded: to a sheeted sub-floor.

A level line is struck on the wall around the floor area at a convenient working height. See Fig. 25. Dots are levelled and squared into this line, and screeds, battens or expansion joints are laid to the dots and at the boundaries to bays of convenient working size. If the floor is to be laid to falls, the dot heights must coincide with the degree of fall and the dot thickness at the lowest point must be at least the minimum thickness for the type of floor. The mix (which should at all times be semi-dry) should be placed in position by shovel and dragged flat, 10 mm above the screed height. The rule should be a heavy floating rule, at least 150 mm wide by 25 mm thick and longer than a full bay width. It is lifted and brought down smartly on to the mix. Care must be taken to see it does not damage the screed or batten.

When the mix has reached the required level it may be either trowelled smooth or flattened by wooden float. At this point the floor is better left and allowed to settle. The

41

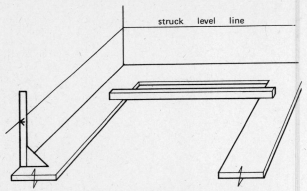

Fig. 25. Ruling in a floor bay using heavy floating rule

finish may be obtained some hours later. If a smooth finish is required, trowel hard at the second trowelling but do not over-trowel. The latter, as well as a wet mix, will result in dusting, because both will cause a skin of neat cement to form over the surface. This will eventually turn to dust.

To make a floor paving of OPC/granite non-slip, carbor-undum chips are lightly trowelled into the surface during the final trowelling.

What is the best way of forming a coved skirting in OPC/sand or OPC/granite?

The best method is to lay the floor first, using battens as screeds or grounds temporarily fixed in the angle of the wall and floor. When the floor has hardened these may be removed. Grounds are then fixed to the required height on the wall surface; the bottom edge of these grounds should be bevelled so that the splay falls away from the wall. This will give a chamfered surface to the top of the skirting.

The tool used to form the coved skirting may be the purpose-made skirting trowel, which should contain sufficient

42

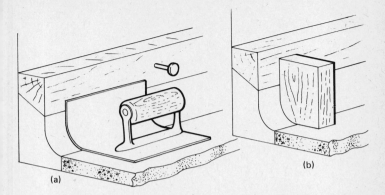

Fig. 26. Coved skirting: (a) skirting trowel, (b) timber skirting gauge

width to form a small upstand: Fig. 26(a). Used in conjunction with this, or on its own if a skirting trowel is not available, is the timber skirting gauge: Fig. 26(b).

How is a lightweight screed applied?

In the same way as an ordinary OPC/sand screed, though it is usually very much thicker, up to 100 mm on average. Once it has been laid and flattened, it should be covered by an OPC/sand screed of 19 mm thickness. This is necessary to protect the weaker lightweight screed. Mixes are: for a lightweight screed, 1 : 6 OPC/vermiculite; for a protecting screed, 1 : 3 OPC/sand.

What method is used to form a staircase in OPC/granite?

The actual laying and forming of the treads and risers must be as for floors and pavings. The sub-floor or base should be saturated and well grouted with a neat-cement slurry.

In general a staircase consists of risers and treads and has either open or closed strings. All treads and risers must be identical in size; to achieve this, riser boards are wedged between walls or string boards at the correct height. See Fig. 27. They

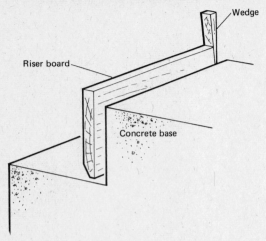

Fig. 27. Forming staircase in OPC/granite

are fixed so that a gap is left between the base and the riser board. OPC/granite is then forced down into the gap until it is full, and the tread is laid in the usual way, being squared in off the riser board behind and ruled from the board in front. The work is tamped and trowelled as in floor laying, with a final trowelling some hours later. The shuttering is struck off next day and all making good is carried out with OPC/granite dust. Later all treads, risers and nosings are rubbed down with a carborundum stone.

What is the 'curing' of OPC/sand renderings, floor screeds and pavings?

To cure means to harden over a period of time. The curing of all cement work is carried out by keeping the rendering or floor damp for as long as practically possible. It may be by sprinkler, damp sacks or covering with polythene.

The applied material is allowed time to harden before the curing takes place, but it must not be allowed to dry out at any point. Two or three days will assist in achieving the final hardness, but for perfection one should cure cement work for up to 28 days.

5

RUN PLAIN MOULDINGS

What is a running mould?

Though very many different types of running mould are used in solid and fibrous plastering, the construction of all of them is fundamentally similar. They all have a metal profile cut *either* to form the required moulding shape *or* to form the reverse so that mouldings may be cast from the resulting mould. The metal profile is backed by a wooden stock. This is cut 3−5 mm smaller than the profile so that when the two are nailed together the metal will project 3−5 mm in front of the stock. The stock is then fixed to a third member, the slipper. In general, the stock and profile are central to the slipper and are supported by struts and braces. Some running moulds also require metal shoes: these are fixed to the major bearing points of the running mould to prevent wear when running *in situ*.

How are metal profiles for running moulds produced?

The profile can easily be determined by first drawing a full-size section through the required moulding in its position *in situ* or on the bench. Taking the outline of the moulding to be the cutting edge of the profile, the rest of the profile is then drawn in.

Fig. 28 shows a section through a moulding on a bench (which could be any other flat background) with the profile drawn over it.

Fig. 29 shows the profile to run the *reverse mould* from which the above moulding will be cast: (a) the moulding is drawn; (b) the required reverse mould, with the appropriate strike-offs, is drawn round it; (c) the profile required is drawn around the reverse mould.

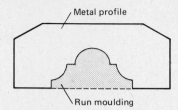

Fig. 28. *Metal profile for run moulding*

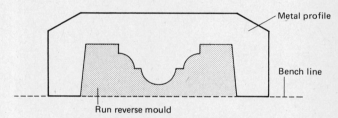

Fig. 29. *Metal profile for run reverse mould*

Fig. 30 shows a cornice in position at the ceiling/wall angle, and also the reverse mould required to cast the cornice. The reverse mould is obtained by drawing in the bench line, as shown, the portions of the ceiling line and wall line enclosed being the strike-offs of the mould. The metal profile is drawn around the reverse mould.

The required section may be marked on the metal by setting out directly with marking tools, or by sticking the paper containing the drawing on to the metal, or by drawing on paper already pasted to metal. The shape is cut out with snips to within 1 mm or so of the line. The rest of the mould is removed by filing to the line with appropriately shaped files. The best cutting edge is achieved by burnishing with a nail or a piece of emery cloth.

47

What is a rebate (rabbet)?

In the context of mouldings run *in situ*, a rebate is a small wooden rule fixed to the underside of a slipper. Its purpose is to rest on the face of the running rule in such a way as to hold

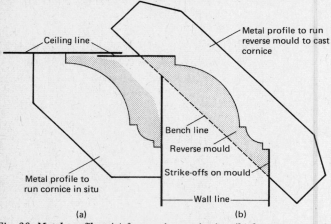

Fig. 30. Metal profiles: (a) for cornice run in situ, (b) for reverse mould
to cast cornice in fibrous plaster

the face of the slipper away from the floating coat or whatever surface is being run upon (see Fig. 32). This eliminates the necessity for a slipper screed.

The term 'rebate' is also used for the step formed around the edges of certain fibrous-plaster casts, to allow for canvas-reinforced stopping; see page 73. This type of rebate is also formed on the ribs used in making a core drum (page 54); the rebates are to receive the laths that form the skeleton of the drum.

What is a muffle?

A muffle assists with the formation of a rough core when running large mouldings in solid or fibrous plaster (see page 54).

It is a false profile formed in either wood or metal cut to a shape similar to that of the profile but usually 6 mm smaller; see Fig. 31. Alternatively it may be formed in plaster that has

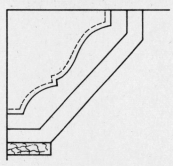

Fig. 31. Dashed line indicates muffle profile

been applied to the running mould in such a way as to project in front of the profile; when hard the plaster is carved to the required shape.

How are mouldings run internally in situ?

By one of three methods (Fig. 32):

(a) run on the finish or setting coat if this is sufficiently hard;
(b) run on screeds or narrow bands of lime putty and class A plaster gauged 1 : 1;
(c) run on a rebate if one surface is either too weak to withstand the pressure or not flat enough to obtain true moulding lines.

Once the decision has been made as to which method one should use, the basic principles for internal mouldings are identical. The running mould must be offered up in position at each end of every run. Nib and slipper points are marked on the backing surface. A dusted-chalk line is then held to these points and a line snapped between the points. Running rules are fixed to the

49

lines and checked at regular intervals for straightness. Large
moulds will probably require a muffle and core run; this will
consist of sand and class A plaster gauged at one part plaster
to three of sand. The core must be well scratched for key.

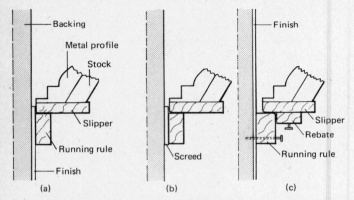

Fig. 32. Moulding run in situ: (a) on finish, (b) on screed, (c) on rebate

The lime putty and plaster should be mixed on the spotboard in equal proportions, and sufficient must be mixed in
the first operation to form one complete length to a good
shape and to rough the second length out. The material is
applied to the moulding area by trowel, and when the length
has been covered the running mould is passed along the entire
length. Rest the slipper on the running rule and push with
the metal profile forming the leading surface of the running
mould. Continue applying more material and running off until
a satisfactory shape has been obtained. Use any surplus material
to force into the background for the next length. A second
gauge will normally be required. This will be a little wetter
than the first but the mix must be the same. Before applying
this, dampen the backing and the rules, then apply by trowel,
forming the shape at the same time. Once more, pass the
running mould along the full length of the run. Two or three

runs will be sufficient to bring a smooth clean finish to the mouldings.

When complete, drop all rules, clean down thoroughly, clear out mitres and prepare to form them.

How does one form a mitre in lime putty and plaster?

The mix must be identical with that used to run the mouldings. The tools required are a gauging trowel, a joint rule, a small tool and a small brush. (See Figs. 46 and 47.)

Apply the material to the area with a gauging trowel or a small tool, shaping to an approximate moulding shape. Then, using the joint rule, which must be at least half as long again as the mitre, form the true shape. This is done by holding the joint rule in both hands; place the right hand fingers on the upper surface, with the thumb underneath, and the left hand fingers under, with the thumb over (for right-handed people). Keep the rule parallel to the moulding members, with the heel down on the finished work and pointing towards the mitre. Draw the rule over the finished work, and the fresh material should form a shape similar to that run. Continue this process with additional mixes until a perfect mitre has been formed. With practice simple mitres may be finished in one gauge, and the more difficult ones in two.

When mouldings are to be run on a bench, how is the bench prepared?

The bench will require a greased running rule fixed down one side. The bench surface should be sealed with shellac and well greased (see below). If the running mould slides over the bench smoothly and without juddering, there is no need to make good the surface and, provided the grease does not dry in, there is no need to shellac the surface. The fewer times a bench is made good and shellac applied, the sounder the top will remain.

The length of the required moulding is marked on the running rule. A nail or nails need to be driven into the bench in the area where the moulding will form, to project but be clear of the running mould. These will anchor the plaster to the bench when it expands on setting and splits itself free. A moulding that is to be taken up will need one nail only, outside the length at the beginning or the end of the run; this will be left behind when the moulding is cut to length and removed. However, a moulding that is to serve as a reverse mould will need nails along its length to hold it in place during casting.

What are the basic stages in making good the surface of a plaster bench?

1. Remove all deposits of plaster by scraping with a joint rule.
2. Clean out all damaged areas.
3. Damp the bare plaster areas.
4. Gauge plaster and make good the surface.
5. Draw excess moisture from the surface.
6. Re-shellac the surface.

How should the damaged areas of a bench surface be treated for making good?

Pitting and scoring on a bench takes place over a period of time, and so all but the newest damage will have been lined with grease for some while. All marks and craters should therefore be scraped and gouged with a chisel. This action will remove a layer of grease-soaked plaster and provide a key for the new plaster. All debris is then swept from the surface prior to damping. The bare plaster areas will require damping to reduce the effect of excessive suction on the new plaster. The amount of water applied should, however, be kept to a minimum since it can weaken the bond of the remaining shellac.

How is the plaster worked during making good?

The plaster is best gauged to the consistency of double cream — as thick as plaster can be gauged for making good — because the strongest possible set is required. Normally, the making good can be achieved with a single bowl of plaster. About one-third of the quantity is tipped on to the bench and pushed over the damaged surface with a joint rule 300—450 mm long, working the plaster well in. The slight suction still afforded by the damaged areas and the agitation of working the plaster will cause it to set. The plaster should all be scraped off as it becomes unworkable. The plaster in the new areas will shrink a little below the surface and be slightly woolly. A further third of the bowl of plaster is worked over in the same way, filling in the areas to near perfection. By this time, the plaster remaining in the bowl will have picked up to a cheesy state and will be ideal for providing the finish to local areas by scraping it across the surface with a joint rule.

Care should be taken that all making good finishes flush with the surrounding areas, and that no film of plaster has gathered on over existing shellac. The adhesion of fresh shellac applied to such a film will only be as good as that of the film of plaster to the shellac beneath it. Such a surface will soon start to flake; therefore any trace of film should be scraped off with a joint rule and the surface rubbed with canvas.

How is shellac prepared for use?

Liquid shellac will merely require thinning to the required consistency by the addition of methylated spirits. Shellac *flakes* are put into a glazed container (contact with any ferrous metal will turn the shellac black) and methylated spirit is then added until it is level with the flakes. The flakes will dissolve in a few days, aided by occasional stirring.

How should a repaired bench be re-shellacked?

Shellac will not stick successfully to over-wet surfaces. Excessive moisture can be drawn from the new areas by sprinkling a 4 mm layer of dry plaster on them and scraping it off with a joint rule some minutes later when it has become damp. All trace of this plaster must be thoroughly scraped from the surface. The shellacking of a bench surface should be kept to a minimum, as an excessive build-up will mar a smooth flat surface and be subject to flaking. The first two coats are, therefore, best confined to the new plaster areas and feathered out carefully at the edges. The third and final coat can then be applied to the whole surface.

How is plasterer's grease made up?

Tallow is melted in a metal container over a low flame. It is then removed to a safe distance from the flame and paraffin stirred in. Although the ratio is around 50 : 50, in the winter a *higher* proportion of paraffin will be needed to produce the correct spreadable consistency and in the summer a *lower* proportion. An alternative to this grease is neat cooking lard.

What are cores and how are they used in running mouldings on the bench?

Cores are the rough in-fillings of large or awkward runs, and may be formed in various ways.

1. Rejected casts and disused run moulds may be positioned in the section to fill it out.

2. Lengths of casts or run moulds may be used, the gaps between them blocked off by draping plaster-soaked canvas over them.

3. The cores may be built up with numerous small pieces of material held in position by a separate gauge of plaster.

4. Vertical upstands can be formed using strips of plasterboard or specially cast plainface.

5. For really large sections, the cores will be drums covered with plaster-soaked canvas and formed by lathing out specially cast ribs, or bricks or building blocks protected by paper for further use.

6. A plaster core may be run up to follow the metal profile accurately some 5 mm smaller, with a muffle fixed over the profile on the running mould. This is known as a *run core* and any of the above cores may be used to core it out. Run cores should be well cross-keyed to receive the finish.

Cores may be used for the following reasons: (a) on smaller sections they save plaster and make it easier to build up the mould; (b) on large sections there is too little time to place sufficient plaster in the correct stage of its set, and also the expansion of a large volume of plaster on setting would cause chattering on smaller sections and stop the mould running altogether on larger ones; (c) they build up awkward sections, such as upstands and overhangs, that would prove impossible to run up in one operation within the set of the plaster.

How is plaster of Paris worked when running a moulding on a bench?

The plaster should be gauged to the consistency of single cream. When a core is being used, the plaster should be worked well into the keying by rubbing with the hand as the plaster is applied and before it has stiffened due to the suction from the core. The core's suction should be used to aid in building out the section. The plaster should be run along the high members first, because these are the most difficult parts of the section to build up and the plaster will always run down, filling the low parts. Large sections, even if well cored, will *chatter* (i.e. develop ridges caused by vibrations of the profile) if not expertly run. Such mouldings can be run up with the suction from a precision run-core before much set has taken place and when the plaster is still quite runny. Such a surface will be woolly but fully built up, and can easily be finished with some applications of fresh plaster. Large, plain sections, such as domes, can be finished with a piece of busk and a fresh gauge of plaster.

A run with no core is built up with the plaster as it becomes stiff enough to stand up and be cut to shape by the running mould. It is therefore placed on the bench in increasing quantities as its setting permits. When the plaster first stiffens, care should be taken when removing it from the gauging container and placing it, because it is a great advantage to have plaster stiffened by the early stage of its set. Handled roughly, such plaster will become runny again and the advantage of an early build up will be lost; plaster thus 'knocked back' will set more quickly. The plaster remaining each time in the container should never be disturbed. As the plaster progresses through its set, however, it will reach a consistency when it can only be used after it has been knocked back. This is done by beating it in the bowl with the hand. Plaster treated in this way will set extremely quickly; it must be placed as soon as it has softened and must have the running mould passed over it immediately. At this stage the section should be fully filled out.

In all but relatively small sections, a small gauge of creamy plaster will be required to provide the best possible finish to the run. This should be gauged and applied as soon as the main gauge has become unworkable, so that the run may be finished before much expansion has taken place. The second gauge does not form a layer over the first but fills the surface to improve its smoothness. It should be taken off as soon as possible after it has been applied, before the suction from the section makes it too stiff. The first application may be rubbed lightly into the surface with the fingers, care being taken not to knock back and so rub away the arrises.

In the later stages of running, workable plaster placed across the beginning of the section will be scraped down its full length by the profile.

In general, the surfaces of mouldings used as reverse moulds need to be strong to withstand the wear and tear of casting. They should therefore be cut from strong plaster that has not been over-weakened by too much knocking back.

How should a running mould be operated when running a moulding on the bench?

Although the bench and running rule are greased, the bearing points of the running mould need greasing to ensure smooth running; these are the underside of the slipper/nib-slipper and the back of the slipper. Greasing the stock helps with cleaning during running and, if the mould is to be cleaned afterwards for further use, it should be greased all over. The mould should have no plaster adhering to it anywhere, since there is always a chance that it will come loose during running and drop on to the run. The running mould is run from right to left and, when run single-handed, is held as in Fig. 33, the right hand at the intersection of the stock and slipper, and the left hand further out along the stock. The action is to walk backwards, pulling the mould smoothly and continuously behind. The arms

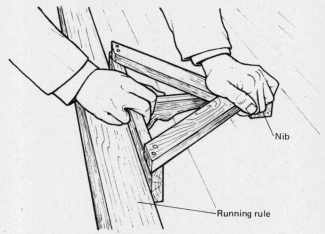

Nib

Running rule

Fig. 33. Running mould being used on bench

are bent to a comfortable position, exerting downward pressure on the mould, and kept rigid. This action causes the running mould to be pulled against the running rule throughout the operation. (If it were pushed ahead, the line of force would cause it to be pushed away from the rule.)

With small moulds, when the left hand is at the nib, care should be taken that only downward pressure is exerted; excessive pulling with the left hand will cause the running mould to pivot about the front of its slipper so that the back end will leave the running rule. This is often done by beginners, causing the profile to cut great hollows along the nearside of the moulding.

When two men are running the moulding, the first takes hold of the mould as above. The second grasps the back end of the slipper (or the strut attached to it) with his right hand only, and pushes continuously by walking forward. As described above, his natural line of force would tend to push the mould away from the rule. He must therefore ensure that he pushes forward and slightly in towards the rule. On larger running moulds there will be room for three or more men to get a hand to it. With wide mouldings, greater difficulty will be found in exerting downward pressure at the nib slipper. It is useful, therefore, to have a man there, pushing down, either by reaching from the other side of the bench or, if the bench is too wide, by walking along the top of the bench and pushing forward. With the largest of moulds, weights (such as bags of plaster) may be used to keep the moulds down on the bench so that the moulding can be cut against the swell of the plaster. The running mould is a framework of braces and, no matter how well constructed, will be influenced by forces acting upon it. The operatives should therefore take hold of the mould in exactly the same way each time and exert exactly the same pressure. If this is not done, in the later stages of running, when the plaster has started to expand, gathering on (see below) and chattering may be brought about.

The running mould should be cleaned after each run so that it is ready to be run again immediately fresh plaster has been applied to the moulding. The surfaces that bear on the bench and running rule must be kept clean to ensure that the mould passes over exactly the same course each time. If the running mould is allowed to lift, a film of plaster will creep under the profile to leave a furry deposit on the surface; this is known as 'gathering on'. The profile must also stand free and plaster is, therefore, cleaned from in front and behind it; the plaster may be wiped from the mould with a piece of lath or washed off with a brush at the slosh tank. In the case of the latter, care should be taken not to allow the water to run on to the moulding from the stock and so weaken the face of the plaster. In the later stages of running a moulding, when the grease has been worn from the bench, a little water may be applied where the slippers run to lubricate the mould. Large sections need to be completed before the plaster expands to any degree, and generally have to be left when chattering occurs. Chattering can sometimes be removed by running the mould once backwards. With sections that tend to expand upwards, one or two more runs may be made after chattering starts to occur. The running mould should be allowed to lift fractionally and ride over the section, thus being required to cut off less plaster. Only experience — to judge the feel of the cutting and vibration of the chattering — can determine to what extent this can be done.

How are curved mouldings run?

Many methods exist for running curved mouldings, but the two that can be used in most instances are:
(a) spinning from a pivot with a gig stick;
(b) running against a curved running rule.
The first method, spinning, is limited to circular curves that are small enough in radius to fit in the space available and are within the bounds of constructing a rigid gig stick. The edge of the gig stick must be flush with the face of the stock in order that the pivot lines up with the metal profile; see Fig. 34. The moulding is run up as for straight mouldings, with the following differences. When running complete circles, the moulding cannot be released from the bench by cutting off the portion containing the nail. Therefore, two nails are driven half into the bench opposite each other and their heads removed. The position of the nails should be marked on the bench outside the path of the moulding so that, when it has set, a thin metal tool may be inserted underneath it at each nail and the moulding levered up.

When spinning, it is easier to build up the moulding in the early stages, as the plaster is applied, by 'feeding' the running mould which passes over the section many times on revolving. In the later stages, however, when the plaster expands as it sets, it can only move outwards as the moulding grows in length and the circle becomes bigger. This will lead to a good

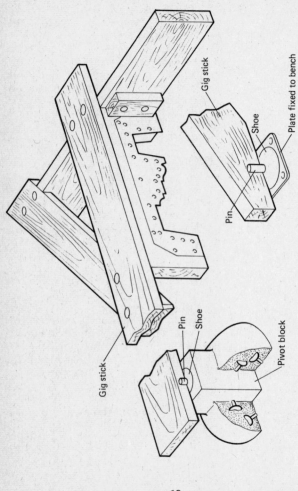

Gig stick

Gig stick

Pin

Shoe

Plate fixed to bench

Pin

Shoe

Pivot block

Fig. 34. Arrangement for spinning a curved moulding

finish on the outer-facing surfaces of the section as they are cut by the profile, but to gathering-on and furring on the inner-facing surfaces. The section must therefore be fully built out and finished with the second gauge, before much expansion has taken place.

In the second method, the running mould used to run against a curved rule is termed a *peg mould*. When the rule is concave, only the ends of the slipper will touch it, enabling the mould to run correctly. Running moulds for convex rules, however, require a block or peg at each end of the slipper (Fig. 35). In

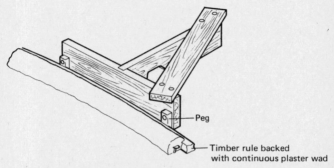

Fig. 35. Running a curved moulding against curved running rule

both cases the running mould must be made before the running rule can be fixed, in order that the rule may be set out parallel with, and the correct distance from, the line of the moulding. The moulding is run up as for straight mouldings.

How does one construct pivots for spinning?

The pin may be a nail driven directly into the running surface: a masonry nail in the case of brickwork and blockwork, or a headless wire nail through a previously fixed square of zinc (to prevent the pin working loose during running) in the case of soft plaster surfaces. A pivot block may also be used; this comprises a pin, which may be a headless screw or nail, in a block of wood wadded to the running surface or fixed to bearers across openings.

6

OPEN MOULDS

What is a flood mould?

It is an open mould made by pouring the moulding material over a model. Retaining walls are erected round the model some 50 mm away, to a height that will allow the material to flood the model until it is submerged. The main moulding materials are polyvinyl chloride (PVC) compound and poly-sulphide cold-pour compound.

What is a PVC skin mould?

It is an open mould made by pouring molten PVC over a model with retaining walls, forming a skin of PVC some 3—5 mm thick over the high points. Skin moulds always require a plaster back (see later) and are used when it would be impractical to flood the model.

What is a cold-pour skin mould?

It is an open mould made by brushing a thixotropic grade of cold pour over a model. When fine detail is to be reproduced, the thixotropic grade can be preceded by a thin coat of standard grade. The required thickness is built up in 3—4 mm layers. These moulds always require a plaster back (see later). Unlike PVC skin moulds, those in cold pour may be formed round undercut sections, in which case the plaster back will need to be made in separate pieces to draw away from the mould.

What is PVC moulding compound?

It is a thermoplastic vinyl resin material, available in various grades of differing flexibility and with melting points from 120 to 170°C. It should be melted in a thermostatically controlled electric copper. The material has the property of exactly reproducing the texture of the surface on to which it is poured. *It is naturally oily and needs no seasoning when casting from it.* The manufacturer's safety instructions should be strictly observed, especially regarding contact with the skin and breathing the fumes during melting. PVC can be melted down and re-used until such time as, on the verge of burning, it becomes too thick to pour.

How should models be seasoned for pouring in PVC?

The material most used for models is plaster of Paris. Because this is porous, blemishes may arise on the face of the mould caused by the contents of the pores expanding on contact with the hot PVC. With varying degrees of success, many treatments are used for plaster models, some of which are: damping the model to the consistency of newly set plaster; shellacking and oiling it; sealing it with a proprietary sealer; or, where possible, pre-warming it to reduce the temperature difference. The three methods the authors have found to give the best results are: completely sealing a dry model with two-pack polyurethane varnish; fairly heavily oiling a damp model, but ensuring that the oil soaks in fully; or soft soaping a damp model.

If stock models are to be taken from a mould, special dense hard plasters may be used.

All other porous materials (brick, stone, etc.) are best treated as for plaster, bearing in mind that PVC will reproduce faithfully the texture of the model. Therefore, if the texture of brick or stone is required, the model should not be made to shine with a sealer.

Non-porous materials, such as clay, composition, plastics, metals, etc., need no preparation and produce perfect surfaces on PVC moulds. Wood is the worst possible material for models and should be avoided at all costs.

What is cold-pour moulding compound?

It is a liquid polysulphide compound which, when mixed with a curing agent, sets to a flexible rubber. The curing time can be varied by the use of differing curatives and it is available in various flexibilities. Once cured it cannot be re-used. It is extremely penetrative, reproducing extremely fine detail and requiring extra care when sealing the joints in retaining walls, etc.

Polysulphide cold-pour compound requires seasoning when casting from it. It should be given a coat of the manufacturer's release agent or of petroleum jelly.

How should models be seasoned for pouring in polysulphide cold pour?

Polysulphide cold pour can be poured on models of clay, damp plaster, plasticine, metal, wood, glass fibre and most plastics. Seasoning is not required, but a thin coat of the manufacturer's release agent or petroleum jelly thinned with an equal quantity of paraffin will aid in parting, and can be used on all but clay and plasticine.

How are the retaining walls for open moulds formed?

Straight walls are most easily formed with strips of wood. These are fixed to one another and to the bench by wads (see page 75), using a nail in the wood and the bench as a key. The joint between walls and bench can be sealed with plaster or clay. Curved walls, or those following a complicated line, may be formed by a strip of zinc or other profile metal or from strips of clay. The metal will require backing up with plaster wads keyed to the bench with nails.

When it is intended that a plaster back be cast down to the bench (see page 67), the walls should slope inwards.

How are PVC open moulds poured?

The PVC must be runny enough to flow and reproduce completely every detail of the model. However, the cooler the material the fewer imperfections are likely to be formed on the face of the mould. For this reason, the buckets of PVC are often left after tapping from the copper until a skin has formed on the top.

The skin is then removed, and pouring takes place from one corner to move smoothly across the model behind the advancing flow. As one bucket is emptied, the next should be ready to continue the flow.

In the case of skin moulds, the PVC is poured over the high points first and the resulting film further thickened by scooping PVC from the low flooded parts with a small tool or gauging trowel, care being taken not to damage the surface of the mould.

How are cold-pour flood moulds poured?

The two constituents should be mixed together thoroughly by stirring (not beating, since this will tend to entrain air). During pouring the mix should be allowed to flow slowly over the rim of the container, causing any entrained air bubbles to burst. Low temperatures will slow the cure.

When do flood moulds require a plaster back for support?

Although the back of a PVC flood mould is perfectly flat immediately after pouring, it seldom remains so on solidifying. The back may become irregular for two reasons: (a) PVC shrinks on solidifying, so the thicker areas tend to shrink and form undulations; (b) eruptions, due to escaping gases, cause the back to become pitted. Such moulds, when turned over, will distort as they settle to fit a flat surface. They therefore require a plaster back as a support.

The backs of cold-pour flood moulds will usually be flat enough not to require a plaster back. Long, narrow moulds, straight or curved, in PVC or cold-pour, will require plaster backs to hold them to line, however.

66

What forms do plaster backs to flood and skin moulds take?

They may be confined to the back of the mould by being cast within the retaining walls, or extended right down to the bench by removal of the walls. In the latter case, the walls should have been splayed inwards to allow the plaster back to draw from the mould. See Figs. 36 and 37.

Backs to small flood moulds will be merely one gauge of plaster containing a couple of layers of canvas. (Details of the use of canvas, wood, etc. to reinforce plaster casts are given in Chapter 7.) The backs to large flood and skin moulds will require to be the strongest form of cast, strong enough to support the PVC and cast and to withstand constant handling. They therefore often contain 25 x 50 mm timbers on ropes around the perimeter and closely spaced in both directions across the back.

How is the back of an open mould prepared before the plaster back is cast?

The external angle, where the material meets the retaining walls, must be removed. This is best done with a sharp knife, held at 45° across the angle to produce a cut surface approximately 15 mm wide. This will help locate the material into a back cast to the walls only; with backs that are continued down to the bench, it will ensure that the material will fit into them. In addition, extra locating Vs can be cut at intervals along the angle. Care must be taken not to disturb the material before the back is cast, as often it will not sit back on the model in its original position. Cold-pour will first require seasoning.

How is a plaster back prepared to house an open mould?

All protuberances formed in the plaster back by irregularities in the PVC must be cleaned off. The whole of the back must then be sealed with shellac, because bare plaster will draw the oils from the pliable compound. The whole is then french chalked prior to greasing. The strike-offs are greased and the

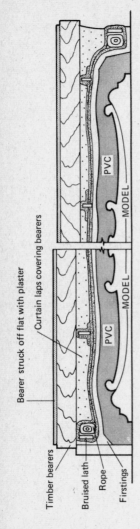

Fig. 36. Section through PVC flood moulds of circular ceiling centre: (a) plaster back stops at retaining wall, (b) plaster back extends down to bench

Bearer struck off flat with plaster

Curtain laps covering bearers

Timber bearers

Bruised lath

Rope

Firstings

PVC

MODEL

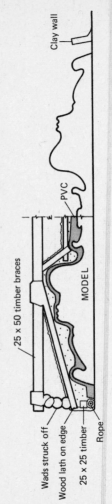

Fig. 37. Section through PVC skin mould: right, model with clay wall ready for pouring; left, fibrous plaster back cast over PVC skin

Clay wall

PVC

MODEL

25 x 50 timber braces

Wads struck off

Wood lath on edge

25 x 25 timber

Rope

grease extended a short way under the pliable material to allow any plaster creeping under to be easily cleaned off.

When do PVC flood moulds require a PVC back?

When a flat back is required on a mould that is to be bent. (This is done either to a curved line on the bench or over a curved surface.)

How is a PVC back to a PVC flood mould formed?

A second pouring is made over a solidified flood mould. Normally, molten PVC does not stick to a previous pour, but the two can usually be prevented from parting by careful handling. It may be found that a very light oiling will act as a flux and help to weld the two pours together.

7

CASTING

What is fibrous plasterwork?

It is plasterwork, containing reinforcement, that has been produced as a cast from a mould. The plaster is usually plaster of
Paris, although harder plasters, such as Keene's cement or
special class As, may be used in certain cases. The reinforcement is usually two layers of jute canvas with strips of wood
lath spaced at a maximum of 300 mm centres. Tow or synthetic
fibres may be used instead of canvas, and metal sections or
heavier timber instead of wood lath.

Other materials may be used by the plasterer to produce
casts, however. Those to be used externally need to be cast
from a cement and aggregate mix and reinforced with a rustproof metal mesh and metal sections. The aggregate is usually
sand, but stone dust is used when producing reconstructed
stone. Composition and gesso enrichment has been used to
dress moulding since the Adam brothers in the second half
of the eighteenth century; probably, had it been available
then, Robert Adam would have used glass-reinforced plastics —
a valuable material for external moulded work.

What is fibrous plastering?

It is the production of fibrous plasterwork. This includes
making moulds — either reverse moulds using running moulds
(Chapter 5) or open moulds using models (Chapter 6); *casting*
the reinforced plasterwork (this chapter); and *fixing and
stopping* (Chapter 8). The craft of fibrous plastering there-

fore includes important processes both before and after the actual casting of fibrous plasterwork.

How is plasterer's size (retarder) made up?

The powdered gelatine-based size is made up into a solution according to the instructions on the packet, usually by soaking and boiling. When the solution is cool, hydrated lime should be added in the proportion of one heaped gauging trowel per bucket to combat infection and prevent the solution from gelling.

How and when is size used in fibrous plastering?

Size water is stirred and a measured quantity added to the gauging water when the plastering operation cannot be carried out within the setting time of the plaster. For correct usage, it is necessary to understand how plaster is worked. For successful *casting*, plaster must be applied in a liquid state and, although it may pick up (due to the agitation involved) during application, the use of size will ensure that plaster yet to be applied remains liquid. When *running and stopping*, the plaster is worked while it is changing from liquid to solid during its set, and so size is not used. If setting plaster that contains size is knocked back, it will be found to be setting unevenly and to contain pips and small, hard lumps. It follows that size can be used when *running a core with a muffle*.

How may the set of plaster of Paris be accelerated?

Potassium alum is made up into a saturated solution with water and added to the gauging water.

How may plaster of Paris be hardened?

Dextrene is made up into a solution with water according to the instructions on the packet. When casting with firstings, a

71

quantity of the solution is added to the gauging water. Although slight retardation of the set is a side-effect, the quantity of size to be added should not be materially affected.

What is the firstings and seconds method of casting?

Two gauges of plaster are used, the *firstings* containing less size than the *seconds*. The firstings, which will form the face of the cast, are applied to the mould and allowed partially to set before the canvas reinforcement is applied to it. The seconds are then used to complete the cast with the remainder of the reinforcement.

What is one-gauge work?

A system of casting using only one gauge of plaster. The layer of plaster applied to the mould is dusted with dry plaster. This stiffens the plaster and prevents the canvas being worked through it to the face.

What are the basic stages in producing a fibrous plaster cast?

1. The wood lath reinforcement is cut and put to soak.
2. The canvas reinforcement is cut.
3. The mould is seasoned when necessary.
4. The plaster, containing an appropriate amount of size retarder, is gauged.
5. The layer of plaster that will form the face of the cast is applied to the mould. This layer may be the firstings in two-gauge work or part of the single mix in one-gauge work.
6. The layers of canvas and strips of wood lath reinforcement are applied with the rest of the plaster.
7. The strike-offs are made up across the back of the cast.
8. When set, the cast is removed from the mould and its face is cleaned up.
Details of typical casts are shown in Figs. 38 and 39.

72

What is the function of rebates on fibrous plaster casts?

A rebate is a step, 25 mm wide and 6 mm deep, formed round the edges of a cast that is to be butt with another cast and is to be stopped in to form a flush joint. It is formed by placing a

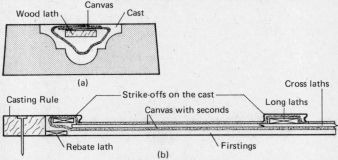

Fig. 38. Sections through casts, showing canvas and lath reinforcement:
(a) moulding, (b) plainface

greased lath or strip of plastic or rubber round the edge of a mould; see Fig. 38. The two rebates come together to form a recess that will house 100 mm plasterboard scrim in the stopping.

What are the functions of the canvas and wood-lath reinforcement?

The chief role of the canvas is to impart tensile strength to the plaster, allowing it to be produced in sheet form. It also binds the plaster round the timber reinforcement, and will hold a cracked or shattered cast together until it can be made good.

The laths form the stiffening ribs across the back of a cast and act as its fixings.

What is the function of 'strike-offs' on the back of fibrous plaster casts?

Strike-offs are screeds of plaster round the perimeter and across the back of casts, and are formed by ruling in from the

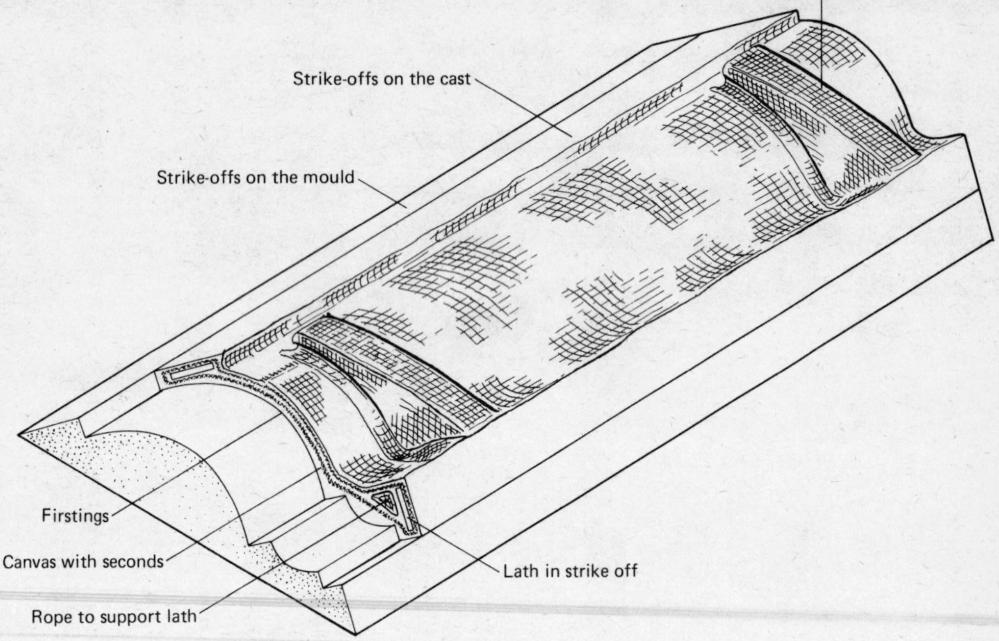

Lath cross-bracket covered with a lap

Strike-offs on the cast

Strike-offs on the mould

Lath in strike off

Firstings

Canvas with seconds

Rope to support lath

Fig. 39. Fibrous plaster cast of cornice in its reverse mould

strike-offs of the mould; see Fig. 39. They should always be on top of the wood lath reinforcement that will be used to fix the cast. They ensure that the face of the cast is in its correct position when the cast is strained back against the fixings.

What are laps and wads?

Laps are small pieces of canvas that are used in two ways: (a) as general canvas reinforcement to cover awkwardly shaped moulds, and may be placed dry or soaked in plaster and opened out; (b) in the form of plaster-soaked strips, opened out to cover wood lath reinforcement in a cast.

Wads subtly differ from laps in that they are rectangles of canvas soaked in plaster but are not used as continuous cover. They are folded into thick strips, approximately 100 mm wide, and used to joint one unit to another, e.g. casts to metal fixings, one cast to another, lifting-out sticks to the backs of casts, etc.

How are laps and wads cut?

Laps are cut from a roll of canvas at the canvas bin by folding the canvas a number of times to resemble a bolt of material and then cutting completely through it lengthwise with a canvas knife. This will provide rectangles of canvas of the same width as the bolt. They may then be cut with the knife to a smaller size by placing them in bulk across the slit in the top of the bin.

On site, when large numbers of wads are required for mass wadding-up, rectangles of canvas are produced by cutting through a roll of canvas lengthwise. These rectangles can be cut to a smaller size by placing them in bulk across the gap between two scaffold boards.

What are the functions of lifting-out sticks?

They are timbers, usually 25 x 50 mm, wadded across the backs of casts and left to project some 150 mm at each end. They serve as (a) handles by which the cast is lifted from the mould and carried around, (b) legs on which the cast is stood or stacked, (c) stiffeners and braces to support a cast that might be subject to breaking during handling. They are cut off on site immediately before fixing.

What are false brackets and cross brackets?

False brackets are pieces of timber reinforcement, usually wood lath, wadded within a cast in order to support fragile sections, such as beamcase sides. They are cut out immediately before fixing.

Cross brackets are pieces of wood lath reinforcement positioned across the width of a relatively narrow cast; see Fig. 39.

What is a rope?

It is a form of reinforcement in a cast, comprising a strip of canvas soaked in plaster and lightly wrung out. Although it may be used instead of wood lath (as cross brackets for small sections), it is generally used in conjunction with wood lath in the following ways: (a) to box laths; (b) placed under flat laths to prevent their cracking the face of the cast; (c) to help build up a deep strike-off; (d) placed over a curved surface under a bruised lath to compensate for the weakening of the lath that bruising produces.

What is a boxed lath?

A rope is placed against the inside edge of a cast. One lath is positioned on edge against the rope and a second placed flat on top of it (Fig. 40). This forms extremely strong edges to a cast, and provides the fixing facilities offered by both flat and on-edge lathing.

Where are wood laths positioned in fibrous plaster casts?

Wood laths should be positioned all round the perimeter of a cast and across it in both directions, spaced at a maximum of

300 mm centres. When a cast is to be nailed or screwed to timber fixings the laths are placed flat, and when the cast is to be fixed by wire and wad to metal fixings the laths are placed on edge; see Chapter 8.

What governs the length of wood laths to fit in a cast?

Wood laths to fit a mould are cut some 20—25 mm shorter than the space they are to occupy. This allows for the thickness of the first layer of plaster and canvas that precedes them.

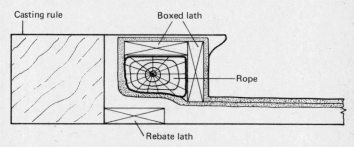

Fig. 40. Section through edge of cast, showing boxed lath

Wood laths to fit between other wood laths, however, need only a tolerance of 1 mm or so.

How is a joint made in long laths in a cast?

The two ends are butted together and reinforced by placing a piece of lath some 150 mm long, acting as a splint, by the side of or on top of the join, depending on the space available.

How are wood laths made to follow curves in a cast?

Wood lath may be made to remain curved by bruising it, using the head of a lath hammer. A short piece of 6 mm lath is placed on the floor and the lath to be bruised is laid at right-angles across it. This upper lath is struck at regular intervals where it bridges between the floor and the under lath. The lath is less likely to break if it is soaked before bruising.

To follow a curved mould with on-edge laths, the laths are cut into short lengths (the quicker the curve the shorter the lengths) and are placed either double in break-joint fashion or singly with overlapping knuckle joints.

Should the wood lath reinforcement of casts be soaked in water prior to use?

Wood laths that are not to be positioned on ropes should be soaked for at least one hour in water to allow them to expand before they are put into a cast. Laths used dry will expand and crack the face of the cast.

How is the canvas reinforcement for a cast cut to cover a mould?

Sufficient lengths of canvas are cut from the roll to cover the entire mould, overlapping it by approximately 100 mm all round. This will allow the canvas to be folded back ('turned in') over the perimeter lath. All joints between pieces of canvas should overlap by approximately 50 mm. The second layer of canvas is cut to cover the mould, but this time finishing just inside the perimeter, lapping the turn-in. Allowance should be made for tucking in around the timber reinforcement.

How are plaster surfaces seasoned for casting in plaster?

1. They are sealed with shellac.
2. On completion, the shellacked surface is dusted with french chalk.
3. The sealed surface is greased before each casting operation, any water left behind on a mould when a newly set cast has been removed being first wiped off.

How is shellac applied to plaster surfaces?

Shellac should be thinned with methylated spirits. The correct consistency will seal the surface completely if applied in three successive coats, resulting in a uniform shine. The first coat, acting as a primer, is thinned more than the other two; however, too much thinning will result in the shellac adhering insufficiently and peeling off with casting. Each coat is applied with a brush, and when dry is rubbed with a canvas pad and brushed to remove any particles adhering, care being taken not to damage the arrises. The surface of newly applied shellac is dusted with french chalk to harden off the surface.

How is grease applied to a surface for casting?

A correctly greased surface shines with a thin, even film but has no grease marks or texture. On a plain surface grease is most efficiently applied with a pad made from canvas, whilst a brush will be needed to reach into the detail of enrichment. Any grease deposit at the bottom of detail and arrises should be removed with a brush.

How is the first layer of plaster applied to the face of a mould?

The plaster, be it one-gauge work or the firstings of two-gauge work, is applied with a splash brush. A wet brush works easily into the plaster but excess water, which will cause a weak patch of plaster on the face of the cast, should first be shaken from the brush. The plaster is brushed into the mould. On plain surfaces this involves no more than pushing it over the surface. On enrichment, however, the plaster will need to be worked into the detail by stippling (or 'punching') in order to avoid air holes in the tips of the detail. Bubbles, which would cause air holes in the arrises of casts, are squashed by forcing the plaster into the quirks of the mould with the fingers ('running the fingers through the arrises'). The mould will then require re-brushing to replace the plaster wiped from each side of the quirk. Over-brushing should be avoided on greased surfaces, since this will rub off the grease. An even layer 5−8 mm thick, depending on the type of cast, is then flicked on with the splash brush. On horizontal surfaces plaster may be shaken lightly from the brush, while on vertical surfaces quite hard flicking will be required. Violent splashing of correctly retarded plaster of Paris firstings will cause the plaster to pick up. On vertical surfaces this will hold up a second layer of the same plaster until the required thickness is built up.

With two-gauge work the splash brush and containers for the firstings are washed out and the plaster wiped from the strike-offs of the mould ('striking off').

The more stiffly gauged plaster of one-gauge work will need dusting with dry plaster, in order to stiffen the layer sufficiently to prevent canvas reinforcement being worked through to the face of the mould. This is done before striking off, since the dusting would dry the grease from the strike-offs. Over-dusting will result in a dry, separating layer and the face may split away from the cast if knocked. A bag for dusting may be made by gathering the corners of a square of canvas.

How is the first layer of canvas applied to a cast?

The layer of canvas reinforcement is applied when the firstings have stiffened sufficiently to prevent it being worked through to the face. The widths of canvas are stretched tight and lowered into position in the cast. They are then pressed into the back of the plaster. This may be done with the hand or, on flat surfaces, with a gauging trowel. The gauging trowel has the advantage of mashing it into the peaks of the splashed texture while not pushing it too near the face at the lower parts. The canvas should be made to follow the contours of the mould and not stretched tight over the features, e.g. each member along a moulding should be pressed in in turn. Badly applied canvas can cause hollows under the firstings of the finished cast which can cave in on handling. This is termed 'shelling'.

A splash brush is used to brush the plaster used for the second stage well into the canvas. This plaster should blend with the firstings, which should not have set but only stiffened sufficiently to receive the canvas. After the perimeter laths are positioned, the canvas is turned in and struck off to ensure that the strike-offs of the cast are down. The turned-in canvas is then brushed in.

How is the second layer of plaster applied to a cast?

The procedure is the same for both one- and two-gauge work. The second layer must combine with the first that has already been worked through the canvas. The strongest possible cast will be produced by casting 'wet', i.e. by applying a surplus of plaster when 'brushing in' or by splashing on a thin layer after brushing in. This will enable the wood lath reinforcement to be 'puddled in' to soft plaster, leaving no hollows under it, and also allow a subsequent layer of canvas to be pressed in and have plaster ooze up through it, giving perfect penetration. When all the wood lath has been positioned and the second layer of canvas applied (see below), the whole back of the cast is brushed in with seconds and either left with the brushed-canvas texture or given a light splash to produce a fine, mottled texture.

How are wood laths put into plaster casts?

Before the laths are positioned they are pasted with plaster, using a splash brush in order to force the plaster into the surface. Flat laths that are not placed on ropes should be 'puddled in' on a line of plaster; this will prevent hollows under the laths. Flat laths running in both directions may all be positioned in one operation and covered by applying the second layer of canvas. Laths placed on edge, however, are positioned after this second layer and covered with plaster-soaked strips of canvas. Unbroken on-edge laths may be placed in one direction only, with the cross laths being cut to fit between them. The cross laths may be placed either at right angles to the long ones or in herringbone fashion. In order to provide more strength when laths need to be placed flat to facilitate fixing, half- or full-width laths may be placed on edge on top of the flat ones.

How is the second layer of canvas applied to a cast?

It is stretched tight and lowered into position over the cast. Always starting from the centre and working out towards either end, pulling in slack from the ends, the canvas is tucked in round any convolution in the shape of the cast or item of reinforcement, so that it follows every contour and forms no pockets or 'cobwebs'.

The back of the cast should be wet and plaster should ooze up through the canvas. The layer is then brushed in and left with a brushed texture or given a light splash.

When the timber reinforcement of casts is such that it cannot be covered by the second layer of canvas, it is positioned after the second layer has been brushed in, covered separately with plaster-soaked strips of canvas, and then brushed in.

How is the back of a cast struck off?

As a general rule, every part of the cast should be 'down' lower than the strike-off line of the mould. Stiffened seconds (seconds either setting or stiffened to the correct consistency) or a gauge of neat plaster is placed on top of the perimeter laths and all laths in the cast that are to be used for fixing. This plaster is then struck off to shape (thickness, angle, etc.) to fit the fixings provided on site. This is done by ruling off the strike-offs on the mould with devices ranging from short pieces of lath to specially constructed gauges.

How are lifting-out sticks fixed to the backs of casts?

They are wadded to the positions of timber reinforcement as the last operation in casting, after the strike-offs have been built up. Wads should never be attached to the back of a cast between the woods, because they would then pull back the panel, wrecking that portion of the cast and rendering the stick useless. Sticks are more easily cut from the cast prior to fixing if a saw can be inserted between cast and stick. For this reason, pieces of wood-lath packing are inserted under the stick on the strike-offs of the mould prior to wadding so that the stick is clear of the cast.

How is a cast cleaned up when it has been removed from a mould?

Casts are more easily cleaned up when new and wet than when the surface has become dry and chalky. All large, plain areas are scraped to remove the surface grease and protruding imperfections. Busk is best used for curves and joint rules for straight surfaces. All 'fat' is then removed by rubbing the surface with a canvas pad. Plain moulded work may have the quirks cleaned out with a joint rule or French plane, and rubbed with canvas. Enriched work has any 'pips' removed from the detail and, where practical, is rubbed with canvas.

8

FIXING AND STOPPING

How should fibrous plaster casts be stored?

Casts should always be stacked face-to-face and back-to-back,
and are better fixed together face-to-face in pairs. Any warping—
caused by shape or structure — in each new wet cast will then
be exactly cancelled out by the other of the pair, and the face
of a cast will not be marked by another's rough back. Cornices
are best boxed together and tied with scrim, while flat casts,
such as plainface, are best cleated. If plainface casts were to
be stacked flat like plasterboards, the minute undulations in
their strike-offs would be greatly magnified throughout the
stack, resulting in considerable distortion of the upper casts.

All casts are best stood up. Lengths of mouldings tied face-
to-face may be laid down on edge, but if laid flat will sag
between the brackets. Casts should be stood as upright as
possible: tied vertically in a stack against a fixed board is ideal.
Careful tying at top and bottom, as well as in the middle, will
help prevent the casts curling. (The end 300 mm or so of all
cast mouldings tend to bend back on drying out, giving a kick
in the line to such casts butted end-to-end.) When necessary,
all casts should be protected from the weather and raised from
the floor on boards to safeguard against damp. Barriers should
be erected around the storage area as a protection from casual
impact.

How are fibrous plaster casts transported to a site?

The best vehicle for this purpose is a furniture van. The casts
should be stacked in the van in the same way as they are when

awaiting use, i.e. tied upright or laid down on edge. They should be securely tied in position or wedged against all movement and padded with pieces of packing.

What operations are involved when fixing fibrous plaster casts on site?

1. Remove any lifting-out sticks or false brackets from the cast, and trim from the strike-offs any selvedge that will prevent the cast from seating properly on its fixings.
2. Check the cast for correct size, check pick ups, cut to mitres and joints.
3. When fixings are to be made across the width of a cast from below, mark across the face the position of the wood-lath reinforcement.
4. Lift the cast into position and hold it there by hand, struts, hangers, cleats or blocks.
5. Fix the cast to the fixings, straightening and lining through when necessary.

How is a cast secured to timber fixings?

A fixing is made every 300 mm along the timber, normally with galvanised nails twice as long as the thickness of the strike-off, but with rustproof screws in situations where vibration must be avoided. The heads are countersunk some 3—5 mm below the surface and made good. See Fig. 41.

How is a cast secured to metal fixings?

A fixing is made every 400 mm along the metal bearer with doubled 16g galvanised tie wire covered by a wad. With normal wood-lath-reinforced casts, a hole is drilled through the face on either side of the on-edge lath for each fixing. The holes should be tight against the lath but spaced along its length some 60—80 mm apart. To countersink the wire, a groove some 6 mm deep is cut between the holes with a chisel, knife, lath-hammer blade, or the tip of a saw. The doubled tie wire is then bent

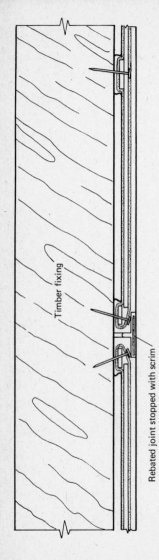

Timber fixing

Rebated joint stopped with scrim

Fig. 41. Section through plainface casts secured by nails to timber fixing

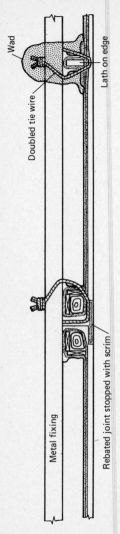

Wad

Doubled tie wire

Lath on edge

Metal fixing

Rebated joint stopped with scrim

Fig. 42. Section through plainface casts secured by wire and wad to metal fixing

into a staple shape, inserted in the holes from underneath and tied from above (Fig. 42, right). When extra fixing brackets have been provided on the back of the cast, the wires may be passed underneath them or through holes drilled in them without penetrating the face of the cast.

How are casts fixed to each other?

1. The simplest type of fixing is where mouldings are held together at mitres and joints by reinforced stopping. The ends of the cast should be well keyed and a gap of 10—15 mm left to house a sizeable wad.
2. A joint across plain areas, curved or flat, may have a 25 x 6 mm rebate on either side so that the casts may be wadded together from the face by reinforcing the stopping with jute plasterboard scrim. This is termed a *rebated joint*.
3. A *lapped joint* is one in which one cast can be made to lap behind another, the underlapping cast being nailed or screwed to the other. Where possible, this form of joint should be rebated on both sides for the stopping and is then termed a lapped and rebated joint.
4. The design of the plasterwork may permit one cast to overlap another, e.g. cornice fixed to plainface areas, or the perimeter mouldings of decorated plainface fixed to the neighbouring plainface.
5. When casts are fixed from behind, a continuous wad, consisting of a double layer of soaked jute canvas, should be placed along the entire joint, lapping round the strike-offs of both casts.

How are fibrous plaster cornices fixed?

Cornices must be fixed to both ceiling and wall. They are fixed to the ceiling:
1. By nails or screws (a) through an existing ceiling into the joists behind; (b) to a timber batten fixed either to a concrete ceiling or across the bearers for expanded metal lath; (c) to

plugs in a concrete ceiling; (d) to the timber reinforcement in a fibrous-plaster ceiling.

2. By wire and wad to metal bearers.

They are fixed to walls and other vertical surfaces (beams, etc):
1. By nails or screws (a) through an existing wall finish into timber studs; (b) to a timber batten fixed to any solid background or across the bearers for expanded metal lath; (c) to plugs in a solid background; (d) to the timber reinforcement of fibrous-plaster units.

2. By wire and wad to metal bearers.

What is the procedure for fixing fibrous plaster cornices?

1. The projection and depth of the cornice are measured and marked on the ceiling and wall, respectively, at the end of each wall.

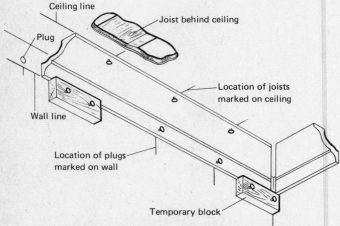

Fig. 43. Fixing cornice to timber ceiling joists and plugged wall

2. The position of the cornice is marked by snapping a chalk line between the marks. When joists behind an existing ceiling

88

are to be used as fixings they are located behind the ceiling line with a pricker and their positions marked with a pencil in front of the ceiling line. When the cornice is to be fixed to the wall with plugs, the holes for the plugs are drilled to coincide with the cornice's wall-line strike-off; the plugs are then inserted and their position marked below the wall line. See Fig. 43.

3. Blocks of wood, on which the cornice will rest during fixing, are fixed to the wall line, usually three to a length of cornice, one at each end and one in the middle. A block should always be placed at the extreme end so that it will support both pieces of cornice at joints and mitres, ensuring that the members line up.

4. The back of the cast is checked prior to fixing, as for all fibrous plaster casts.

5. The wall is measured and the cornice cut to length. This may involve cutting mitres and square joints, and placing them to suit enrichment repeats. See below.

6. The cornice is then offered into position and held there by hand, T-strut or cleat, while it is fixed as previously described.

7. All joints, mitres and fixing holes are stopped in.

How are mouldings cut for mitres?

Two methods are needed: one for mouldings on single backgrounds and another for mouldings on two different planes, such as a cornice to a ceiling and a wall.

In the case of the former, the position of the moulding is marked on the background as part of the setting out. The moulding is offered on to the setting out in each position in turn, and its width marked on to the background. All the mitre lines may now be drawn in on the setting out by drawing through the intersections (Fig. 44). Each piece of moulding may then be offered into its position and the mitre cut determined by marking the mitre line on both sides of the moulding. Right-angle mitres will have a mitre line at an angle of $45°$, so this can be marked directly on the moulding with a $45°$ set square or be cut with the use of a mitre box.

In the case of a cornice, to obtain the correct mitre cut the moulding must be held in its correct projection and depth and be cut by a saw held plumb. This is achieved by using the bench or working surface as the ceiling, and holding the cornice with

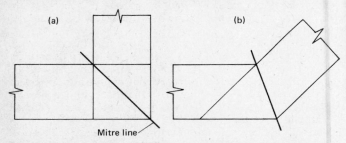

Fig. 44. Mitre lines for mouldings intersecting at (a) right angle, (b) obtuse angle

its ceiling strike-off flat on the bench. (This should bring the wall line of the cornice to the depth of the cornice above the bench.) If many mitres are to be cut, a board representing the wall may be fixed perpendicular to the bench to make holding the cornice in position easier. The depth and projection of the cornice may be marked on the board and bench, respectively, and a batten fixed to the bench to the projection line will lock the cornice in position (Fig. 45).

The mitre is set out by first marking a line square across the moulding from the wall line. The projection of the mitre is then marked from the square line along the ceiling line, outside the square line for an external mitre and inside for an internal mitre.

With right-angle mitres the projection of the mitre will be the projection of the cornice. With acute and obtuse mitres the projection of the mitre is obtained by setting out the projection of the cornice on the ceiling at both sides of the mitre, and then measuring back from the mitre line to a square line marked on the ceiling from the wall angle.

With large mouldings an attempt should be made to mark

the mitre line over the members, using a square. But on smaller ones it is sufficient to hold the saw plumb, sighting it through the marks on each side of the moulding.

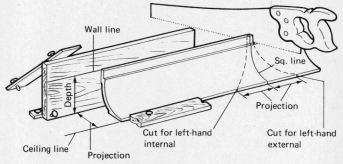

Fig. 45. Cutting a mitre in a cornice

When marking the length of a moulding, the moulding should be cut shorter to allow material to be placed in the gap at each mitre and joint (see next question).

What constitutes sound stopping and making good?

Stopping to fixing-holes (nails, screws, wires) does not need reinforcing. However, tight fixing of the cast is essential because any subsequent movement may result in the stopping's being pushed off by nail or screw heads. Joints between casts or between a cast and its background are subject to movement, and the stopping between them should contain jute scrim reinforcement. Mitres and joints in mouldings should have a wad placed in a 10–20 mm wide gap left between them, a wad being infinitely superior to a few canvas strings pulled from the weave. The stopping to rebated joints should contain a layer of plasterboard jute scrim and the rebates should be well keyed. The joint where a cast meets a background (the wall and ceiling line of a cornice, etc.) should contain a string of canvas, even if room has to be made for it by a saw cut, etc.

91

Sound, strong plaster should be used. Dead plaster will have no strength and will fall out due to its not adhering to the casts and not expanding on setting. Joints are sounder if they are brought out in one gauge. If the wads are put in first and the joint finished with another layer of plaster, there is obviously the possibility of separation. Stopping and making good should always meet a clean edge, because plaster cannot be feathered out. All hollows should therefore be cut out to receive the stopping and well keyed.

How should fibrous plasterwork be keyed?

With a sharp chisel to form undercut grooves, varying from 20—30 mm apart and 2—5 mm deep.

What is the sequence of operations in stopping joints and mitres?

Before stopping, the canvas is cut and the joint checked for alignment. Any major adjustment, e.g. releasing the fixing and realigning, is carried out before gauging the plaster. The plaster is gauged in the stopping bowl to the consistency of single cream. So that the plaster is not disturbed once it has started to set, the canvas is put in to soak immediately. While the plaster is setting to a non-drip consistency, the joint is damped until it has the water content of newly set plaster (a little suction aids stopping) and all minor preparation, such as cutting edges to hollows and keying, is carried out. When stopping-in on the bench, where gravity assists, newly gauged, runny plaster may be poured into the joint, but the plaster cannot be cut to shape until it has partially set. When working against gravity, however, the plaster needs to be partially set: runny plaster will drain from the wad, leaving it weak and unable to adhere to the joint; it is impossible to place plaster without body. The joint is built out and a finish achieved by repeatedly placing plaster as it progresses through its stages of setting. It is applied with a gauging trowel or a trowel-end small tool (Fig. 46), just a little full of the section, and then cut off with a joint rule. Stopping to plain areas may be polished when set with a piece of busk lubricated with a little water.

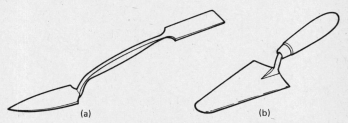

(a)   (b)

Fig. 46 Stopping tools: (a) trowel-end small tool, (b) gauging trowel

How is plaster of Paris worked during stopping?

To achieve a satisfactory finish and maximum efficiency (e.g. optimum quantity of stopping per gauge), the plaster needs to be gauged to one particular consistency. If the plaster is gauged too thickly, i.e. dough-like, so that it may be placed against gravity straight away, it will set too quickly and will produce a rough, woolly texture when cut with the joint rule. The thinner the plaster is gauged the longer it will take to set, but too thin a plaster will be too weak. The desirable consistency is that of single cream; plaster so gauged will cut smooth and clean as it stiffens through its stages of set. It is ready to use as soon as the bowl can be tipped up without the plaster flowing. On plain surfaces plaster is repeatedly placed a little proud of the surface with a small tool or a gauging trowel and scraped off with a joint rule until, at the end of the set, the finish is achieved. The canvas for the stopping should be confined to one part of the bowl in the smallest area possible so that the maximum amount of plaster remains undisturbed. The plaster will settle slightly during the five minutes or so it takes to reach the right consistency. Plaster should therefore be taken from the centre and bottom of the bowl, thus using the thickest first. The agitation of the plaster as it is used will

93

accelerate its set, so it is essential not to disturb the plaster remaining in the bowl. Working outwards from the centre for each application, the deposit around the sides of the bowl will be left till last. On moulded sections the plaster is cut off one member at a time, working the joint rule into the arrises, rather than scraping outwards and so dragging off the edges.

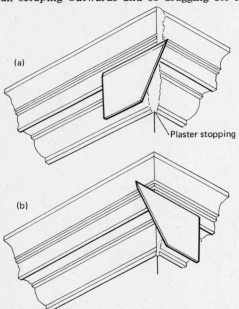

Fig. 47. Joint rule being worked down left-hand and right-hand sides of external mitre

In the case of mitres, the plaster is first removed (a member at a time) from one side of the mitre, and then from the other; see Fig. 47. All finishes should be achieved by cutting and/or scraping plaster in the final stage of its set and not with a brush or by the application of water (unless a special texture is called for).

94

9

FAULTS AND REPAIRS

What faults should one look for in plastering?

Faults may be listed under two headings. First, there are those
that are usually due to poor workmanship. Undulating surfaces,
crooked angles, work generally out of plumb and out of square,
poor-quality finish and generally a poor standard all round. The
best remedy of all of these faults is to have the work done
properly in the first place, by good craftmen, using correct
materials and methods. Good supervision is also very necessary
with large contracts, as is site organisation.

The second series of faults found in plastering are not so
easily defined, neither can the blame be so readily apportioned.
Many of them are due entirely to constructional faults and
defects that eventually appear on the plaster surface. Some may
be due to poor application, others are very often due to faulty
material being used, wrong specifications for related tasks,
and incorrect background or backing preparation. Most of
this type of fault could be prevented but this does not stop
them appearing. The main ones are: blisters and blubbing,
bleeding, blowing, bond failure, case hardening, cracks, crazing,
dusting, efflorescence, flaking, grinning, pattern staining, and
faults in the set.

What faults are found in plastering when using lime?

1. *Popping and pitting*. Due to incorrect slaking, particles of
unslaked lime may be present in the mix. If present in finished
plasterwork, these will expand when they eventually slake,

95

forming craters as they push off pieces of plasterwork. There is no remedy but removal and replastering.

2. *Lime bloom*. This is a film of calcium carbonate on the face of the plasterwork; it is caused by lime dissolved in the water, brought to the face on drying, and reacting with carbon dioxide in the air. Without using lime, this can also occur in Portland-cement mixes, due to the smaller amount of 'free lime' formed in the cement during manufacture.

What are blisters and blubbing?

These are small convex swellings that will appear on the surface of plasterwork. In the main they are due to finishing coats being applied to dusty backing coats. The dust acts as a release agent and small areas of finishing lose adhesion. Dry brushing and then dampening the backing will prevent this, but when it is visible on finished work all blisters should be cut out, the area brushed, dampened and made good with the correct plaster.

What is bleeding?

In solid plastering this fault occurs when water drains from a mix that has been applied but has not had time to harden. This will result in cement mixes not reaching their expected degree of hardness. It occurs usually when a cement and sand or water-proofed mix is being applied to a background or backing that is strong and dense. One remedy is to use a proprietary brand of plasticiser in the mix; this will mean that less water is added, therefore less bleeding will take place. Another remedy when applying waterproofing in several coats is to try to catch the 'green' suction with subsequent coats.

What is blowing?

The pushing off of small portions of applied finishing plaster that contains lime putty. It is due to expansion taking place in small pieces of unslaked lime. It may also be caused by metal rusting (nail heads and the like) or a small percentage of clay being present in the sand. The obvious remedies are to use well slaked lime, to treat all raw metal over which plaster is going to be applied, and to use only clean sand. If the small holes still appear, cut out the cause and make good.

What is case hardening?

This will occur during the application of an OPC/sand mix. It is the hardening of the surface before the mix beneath the surface hardens. Generally badly applied heat will start this off, and it will be made worse by continually working the surface in an attempt to remedy it. The latter will probably result in the surface breaking up and flaking. To prevent case hardening, keep direct heat away from all plaster surfaces, and let any applied heat be gradual. Where it does occur, leave for a period to allow the backing to harden, then work back with either a wood or sponge float and keep the surface damp. This may not be successful if complete dry-out has taken place, in which case the best remedy is to scale the case-hardened skin off and apply fresh material.

What causes cracks in solid plastering?

Large cracks are not often the fault of the plasterer. They are much more likely to be caused by movement within the structure, which will cause the background to split; this will then be transmitted to the plaster surface. The cracks may be due to the settlement of the building, differing rates of expansion between different constructional components, or excessive vibration due to other craftsmen working upon another part of the structure. Where they occur the plaster must be cut right back to the background on either side of the crack, and where possible made undercut. The exposed area must then be brushed dry, free from all loose particles, dampened and treated with PVA. The backing coat should be similar to that used in the first place, and it may be ruled flat from either side of the crack, then cut back to allow for the finish. This is applied in the normal manner.

Smaller cracks may be due to poor application, strong dense mixes used over weaker backgrounds and backings, and shrinkage. These must be cut out, brushed, dampened and (as they are usually small) finished in one-coat finish.

What causes cracks in floor screeds and pavings?

These may be due to movement within the structure. They can also be caused by insufficient attention being paid to the preparation of the subfloor. Where a floor is not thoroughly cleaned dry prior to dampening, the dust will combine with the water to provide an unwanted release agent several months after the floor has been laid. Laying a floor on a dried grout will cause a similar fault to occur. Laying floor bays that are too large will inevitably result in cracking at joints and in the central part of each bay.

Repairing cracked floors is a difficult task to complete satisfactorily. Where the remainder of the floor is sound, the crack may be cleaned out, dampened and refilled with an OPC/washed-sand mix of 1 : 2. The edges may be treated with a PVA adhesive and some of this included in the mix. (This applies to internal floors only.) Where part of the floor has lifted, care must be taken to remove only the loose area; do not hammer indiscriminately as this will loosen areas not affected. Make good regular areas in the same way that a new floor would be laid. Clean dry, dampen, grout base and edges, lay and tamp.

What causes cracks in fibrous plastering?

These are usually due to the use of dry timber or laths as reinforcement to fibrous plaster casts. Cracking occurs immediately after casting, and will be made good during the fixing and stopping operation. To prevent cracking, soak all timbers well before casting commences. Other and possibly larger cracks that appear on finished fibrous plaster are due either to structural movement or incorrect fixing. The latter will cause some movement to take place. All repairs are the same as when fixing and stopping.

98

What is crazing?

The appearance on a finished plaster surface of a number of fine hair cracks in a regular pattern. It may be due to the shrinking of an undercoat after application of the finish (probably caused by using poor, clayey sand in OPC/sand mixes), or to the finish being applied before the backing had set. Excessive untreated suction from a backing coat will also cause crazing, because it will absorb the water from the finishing coat before it has been properly worked. Another cause of crazing is poor application; this means that insufficient pressure was applied to the finish as it was worked. When lime plasters were used more generally, crazing was often in evidence but could have been eliminated by a vigorous scouring with crossgrain float and water before the final trowelling. Once fire cracks, hair cracks or crazing have appeared on a finished surface the whole patch should be scaled off or new plaster applied.

What is dusting?

Dusting is a fault most frequently found in floor screeds and pavings. It may also appear on the surface of cement finishes, in which case it has usually been caused by overworking the surface to obtain a finish, or by drying out too quickly. With floors there are several causes; once again overtrowelling is probably the most common. This will bring a thin layer of neat Portland cement to the surface, and with wear this will turn to dust. Too much water in the mix will cause a similar effect: water and cement will rise to the surface, and will soon turn to dust. Poor sand and a poor mix generally can both cause dusting to occur, as can dead cement and genuine wear and tear. Patent floor hardeners added to the mix or applied to the green floor after laying will help to stop this occurring, as will normal water curing for a period of 7–14 days. One method of carrying out repairs is to apply a self-levelling compound over the area, complying with the manufacturer's requirements.

What is efflorescence?

This will appear on the surface of finished plastering in the form of a white frothy deposit. It may be caused through salt being present in either the background or the backing coat. It may also be caused by excessive dampness due to either a faulty damp-proof course or the lack of one. Usually it will disappear as and when the structure dries out, but may return whenever the dampness does. It will cause damage to various forms of decoration. If time allows, the area should be left till the salts expend their energies.

What is flaking?

It is peeling, scaling or shaling: the breakdown of adhesion between the floating and finishing coats. This will cause areas of setting to break away, leaving the floating coat exposed. It is the result of lack of key, a dusty backing-coat surface, incorrect specification and poor sand or poor mixes generally. Once again the dust will combine with water and form a release agent. Faulty specification may result in a hard finish such as class B gypsum plaster being applied to a soft, friable backing coat. The only remedy is to hack down all affected areas, brush the backing coat and check its hardness; if it is dusting and friable, remove right back to the background. Make good to the level of the floating with a 1 : 3 class A plaster/sand mix, and set in the required finishing plaster.

What is grinning?

This is the appearance on the surface of finished plastering of either the joints in the background or the keying marks of the backing coat. In both cases there is a variety of causes, e.g. suction variation between blocks or bricks and joints, varying background materials, and wrong keying for plaster finishes. However, all could be eliminated if the correct thickness of the various coats of plaster had been applied in the first instance. There is no actual remedy other than to replace with correctly applied backing and finishing coats.

100

What is pattern staining?

This is a fault that is frequently confused with grinning, since it can cause a similar effect under certain conditions. It is, however, atmospheric dust, deposited to an extent that depends on the degree of thermal conductivity: dust settles in greater quantity where the surface is coldest. For example, a ceiling will usually be warmer on the underside than it is on the upper surface between the joists. The better-insulated areas, such as where the wood joists occur, will appear on the lower surface as lighter areas than the less-insulated sections between the joists. In old buildings, dirt on the upper surface may, by permeation, make this more pronounced. Cleaning and lagging will assist in the elimination of this fault.

What causes plaster to set too fast?

When the material sets too fast to allow for normal working, the cause may be one of several. The plaster may have been too hot or fresh, the mixing water dirty, or the plaster stored under damp conditions. With plaster and sand mixes, dirty sand will cause an accelerated set. Dirty tools, appliances, mixing boards, buckets, etc., will cause the plaster to set quickly in small quantities, with the result that the mix is lumpy or knotty. The remedy for all of these faults is self-evident.

What causes plaster to set too slowly?

Plaster that sets too slowly is often referred to as 'dead' plaster. The cause is usually too long storage of the plaster, plaster bags being left open and bags not being used in rotation. Plasters that can be affected in this way will generally have a date of manufacture on the bag. From this it should be possible to ensure that in general no plaster is over three months old. Therefore the storage of plaster is important. It should be dry, off the floor, in a ventilated area, and used in rotation. Where possible, store the bags in an upright position; this lessens the compacting.

How can repairs to plasterboard ceilings (one-coat finish) be carried out?

The damaged area must be cut back to a regular shape and size, such that it will be possible to obtain maximum fixings for the new piece of plasterboard. Cut the old board down the centre of existing wood joists, being careful not to dislodge any of the old fixings. Where this is not possible, either nail additional timbers to the sides of the joists, making extra width, or wedge and nail extra timbers across gaps. At all events, the new fixings must comply in all ways with the plasterboard fixing regulations. All shelling plaster must be removed, and if possible all sound plaster joints must be coated with poly-vinylacetate (PVA) adhesive. Where joints occur in the new board, surface scrim should be applied and then the extra area laid in, ruled off flat to the existing surfaces. Care must be taken that all plaster joints are kept flat and clean. The plaster should be finished as in setting.

What is the best way to make patches in old internal plastering?

The patches must be cut to regular shapes, the edges and the background thoroughly cleaned out, and all loose particles of old plaster removed. The background should then be damped and the edges to all existing work treated with a diluted PVA adhesive. For the backing a 1 : 3 mix of class A plaster and clean sand is best, since it sets fairly quickly and is ready for finishing immediately it has set. Apply this with a gauging trowel, push well into all gaps and edges, and rule flat 2–3 mm behind the finished surface of the old work.

The finishing coat may be a plaster to match the existing work, or it may be class A plaster and lime putty mixed to equal proportions. This should be applied to a full thickness, ruled off flat to the existing work and flush to the edges. It may be either float scoured or trowelled, and finally smoothed and checked by a long straight steel joint-rule for flatness. No joint between new and old work should be apparent to the touch.

How should repairs to existing external plastering be carried out?

Generally there are three stages. First the preparation: clean away all loose particles of the old work and try to make the area into a regular shape; dry brush, dampen and spatterdash. The second stage is the application and levelling of the backing coat. To obtain the correct level or depth of this coat a making-good rule is required (Fig. 48). This is a piece of timber 100 mm

Fig. 48. Use of making-good rule

longer than the gap to be made good. At least 50 mm at each end should be cut back to form two shoulders that will rest on the existing work, so that the main rule will cut the backing coat back to the required depth. Alternatively a smaller piece of timber may be fixed to the first piece so that its face will project to the required depth. When the backing coat has been ruled off and the edges cleared out, it should be keyed and left. The final stage is the finishing coat, which when applied must match the existing finish, with the joints kept clean.

Care should be taken that the materials match in texture and mix. Where the finish is one that did not require keying, or where the thickness is small, as in the resin finishes, a making-good rule may not be necessary.

103

How are cracks made good in moulded work?

In fibrous plasterwork they are cut out and dusted, damped to control the suction, and stopped in with plaster of Paris as for new work. In plain mouldings, the new plaster is ruled in with a joint rule or a piece of busk for straight or curved work, respectively. In enriched work, cracks are filled as the plaster becomes cheesey and modelled in with the appropriate small tools.

In old *in situ* work cracks are cut out, preferably undercut, and dusted to remove all loose material. Although the crack may be damped to control the suction, it is best treated with PVA solution. This, mixed with three to four parts water and painted on in two coats, allowing each to dry, has the added advantage of penetrating friable, sandy undercoats and stabilising them, as well as improving adhesion and controlling the suction.

How may portions of missing plain-moulded work be replaced?

With straight mouldings, the length of the repair will determine the method.
1. Those within the length of a joint rule will merely be ruled in off both sides by working the joint rule across the gap as for normal stopping.
2. Those within the length of a featheredge rule may be worked in, a member at a time, and finished with a joint rule and a separate gauge of plaster.
3. Longer lengths will require a running mould: (a) to run up the section *in situ*, in which case the profile is fitted over the moulding and the running rule fixed to touch the slipper; (b) to run down the section, reinforced with canvas, bed it in position, and make good; (c) to run a reverse mould and cast the moulding in order to fix and stop as for fibrous plasterwork.

With curved mouldings, gaps that are small enough may be wiped in, using pieces of busk, but larger repairs need to be run to the correct curve, either *in situ* or as fibrous plastering.

To maintain the character of aged plasterwork, the running

must be done *in situ* and perhaps with a shortened slipper. This ensures that the moulding follows the undulating backgrounds.

How may portions of missing enriched work be replaced?

A mould will need to be made of a repeat of an existing piece of the enrichment. Casts produced from the mould are fixed and stopped in. The mould may be cast *in situ* as a squeeze, or a section of the enrichment may be removed and taken to a fibrous plasterwork shop for moulding.

Should the paint be removed from enriched plasterwork before a mould is taken?

Yes, in order that the new work will match the original if the whole is stripped of paint in the future. If the rest of the old work is not to be stripped at the time of the repair, the new work can be filled with a water-based paint to match it.

Old water-based paints can be removed with hot water. This softens the paint sufficiently for it to be removed by a combination of brushing and the use of any suitably shaped tool. *In situ*, water is brushed on from a bowl; portions of plasterwork that have been taken down can be simmered in a tray or bucket. Should paints be encountered that cannot be removed by this method, conventional paint strippers can be used.

How may the section of an existing moulding be obtained?

1. By tracing round the section. On the bench this is done by holding the paper against a square-cut end. *In situ*, a damaged end can be cut square or, if no moulding is missing, the paper can be inserted into a saw cut made square across the section.
2. By taking an impression of the section and tracing round the impression. The impression takes the form of a strip of reinforced plaster taken square across the moulding as a

plaster squeeze mould. To produce an accurate tracing, the squeeze will have to be cut square at one end with a saw.

What materials may conveniently be used for in-situ squeeze moulds?

Where no undercut exists, a single-piece plaster cast or clay squeeze mould may be taken. Undercut enrichment, however, will require a skin mould of thixotropic cold-pour supported by a plaster back; see Chapter 6. A clay squeeze may be taken of enrichment containing minor undercut, provided that the undercut is carved back on the plaster cast taken from the squeeze.

How are in-situ clay squeeze moulds taken?

The face of the modelling clay must first be made perfectly smooth and without blemish by pressing it against a smooth surface — glass, plastics, metal, etc. It is then pressed over the model. Surfaces to which the clay is not intended to stick should first be dusted with french chalk. When the clay takes the form of a solid block, a board may be pressed on to its back to support the squeeze during casting. If it is impracticable to make the back of the clay flat, it must be supported by a plaster back. This is formed in the same way as a plaster squeeze. If the plaster is not to stick to the clay, the latter should be smoothed and wiped over with paraffin as a release agent.

How are one-piece in-situ plaster squeeze moulds taken?

They follow the lines of conventional plaster casts (Chapter 7), but are made crude by the fact that gravity does not aid in the application of the materials. First, all the materials are cut and made ready and the model seasoned for casting in plaster. Firstings are gauged thicker than usual and this, together with extra agitation and necessarily slower application, calls for more size than would be used for a conventional cast.

The reinforced seconds are applied in the form of pre-soaked plaster laps. Application is easier if these are in a semi-set state. If the plaster squeeze will fall from the model as it expands on setting, it should be held in position by a strut from below.

Some large plaster squeezes, requiring timber reinforcement, will not remain in position after the firstings have set. The timber should, therefore, be covered in plaster-soaked canvas ropes and strutted against the firstings before its final set. So held in position, the squeeze mould can be completed by applying the canvas-reinforced seconds between and over the woods, as described above.

How are cold-pour skin moulds taken as in-situ squeezes?

The model is seasoned and the cold-pour applied as for conventional skin mouldings (Chapter 6). The reinforced plaster back to the skin is applied as to a plaster squeeze mould.

INDEX

Accelerators, 1, 71, 94
Acoustic plastering, 25–26
Airholes, 79
Alum, 71
Angles,
    external, 21–24
    internal, 20–21
Anhydrous plasters, 2
Arches, solid, 28–30

Backgrounds, 13–14
    plasters for, 8–9
Backing coat, 14–15
Backs for open moulds, 63, 65, 66–69
Bell cast, 29, 30
Benches, making goods, 51–54
Bleeding, 96
Blisters and blubbing, 96
Blockwork, 9, 13
Blowing, 96
Board-finish plaster, 9, 35, 36
Bonding plaster, 9
Boxed laths, 76
Brackets, 76
Brickwork, 9, 13
Browning plaster, 9
Bulking, 7
Butter coat, 6, 31

Canvas, 73, 75, 76, 78, 80, 81
Case hardening, 97

Casting, 70
Casting plaster, see Plaster of Paris
Casts,
    cleaning, 83
    fixing, 85–87
    handling, 84–85
    of a cornice, 74
    of a fibrous-plaster plainface, 73
    of a panel moulding, 73
Ceilings, 35–36
    repairs to, 102
Cements, 2–4
    storage, 8
Chattering, 55, 58, 59
Clay tiles and pots, 13
Cleaning of casts, 83
Coarse stuff, 5
'Cobwebs', 82
Cold-pour compound, 65
    flood mould, 66
    seasoning models for pouring in, 65
    skin mould, 63
    squeeze mould, 107
Concrete, 9
Cores, 54–55
Cork, 13
Cornices,
    fibrous plaster cast, 74
    fixing, 87–89
    mitre cutting, 90–91
    running, 47–48

Cracks,
    in fibrous plastering, 98
    in floors, 98
    in solid plastering, 97–98
    making good, 104
Crazing, 99
Curing, 3, 44–45
Curved mouldings, running,
    59–61
Curved surfaces, solid, 26–28

Dashed finishes, 19, 31–32
Dextrene, 71
Dots, 16–17, 27
Dry dashing, 31–32
Dry lining, 36–38
Dusting, 99

Efflorescence, 100
Enriched work, repairing, 105

Faults, 95 et seq.
    in casting, 79–82
    in running, 55–61
Fibrous plastering, definition,
    70–71
Finishing coats, 13
    external, 30–34
    internal, 18–20, 34
Fire, effect on plastering, 9
Firstings, 72
Fish-tail shoe, 26–27
Fixing,
    cornices, 87–89
    plaster casts, 85–87
Flaking, 100
Floating coat, 13, 14–18
    curved backgrounds, 27
Floats,
    crossgrain, 21
    devil, 15
    skimming, 20

Flood moulds, 63
    backs for, 66–69
    pouring of, 66
Floor screeds, 40–42, 43
    cracks in, 98
    curing, 44–45
    dusting of, 99

Gathering on, 58
Gauge box, 7
Gauging, 7–8
Gigsticks, 26–29, 59–60
Grease, 51, 54, 56, 78, 79
Grinning, 100
Gypsum plasters, 1–2, 6
    gauging, 8
    set of, 1, 101
    storage, 8
    suitability for backgrounds, 8

Hardeners, 71–72
Hawk, 10–12
Hemihydrate plasters, 2
Hydrophobic cement, 4

Insulation,
    acoustic, 25–26
    thermal, 38–39

Jointer, 30

Keene's cement, 2, 9, 22, 70
Keying,
    of backing coats, 15
    of fibrous plastering, 92
Knocking back, 56

Lapped joints, 87
Laps, 75
Laths, 13, 14
    see also Plasterboards, Wood
        laths

Level, 17
Lifting-out sticks, 75—76, 82
Lime, 4—6
    faults when using, 95—96
    storage, 8
Lime bloom, 96
Lime putty, 5, 51

Making good,
    benches, 52—53
    fibrous plasterwork, 91—92
Masonry cement, 3
Metal angle bead, 22—24
Metal furring, 37—38
Metal-lathing plaster, 9
Metal profiles, running moulds,
    46—48
Mitring, 51, 89—92
Models, seasoning, 64—65
Mouldings, run,
    cores for, 54—55
    curved, 59—61
    internal, *in situ*, 6, 49—51
    on a bench, 51—52, 55—62
Moulds,
    open, 63 et seq.
    reverse, 46—47, 56
    running, 46 et seq.
    squeeze, 105—107
Muffles, 48—49

Newtonite waterproofing lath, 9

One-gauge work, 72
Open moulds, 63 et seq.
Ordinary Portland cement, 2—4
    curing, 44—45

Paint removal, 105
Patching, 102

Pattern staining, 101
Pavings, 40, 41
    cracks in, 98
    curing, 44
    dusting of, 99
Pebbledash, 31—32
Peg moulds, 61
Perlite, 8, 9
Pivots, 61—62
Plainface,
    cement, 30
    fibrous plaster, 73, 86
Plaster of Paris, 1—2, 6
    accelerators for, 71
    gauging, 8
    hardeners for, 71—72
    seasoning of, 64, 78
    size for, 71
    working in casting, 70, 72, 79,
        81—82
    working in running, 55—56
    working in stopping, 93—94
Plasterboards, 35 et seq.
    repairs to, 102
    storage, 8
Plasters, 1—2, 19
    lightweight premixed, 8—9
Plumbing dots, 16
Polyvinylacetate, 9, 13
Polyvinylchloride,
    backs, 69
    moudling compound, 64
    pouring, 66
    seasoning models for pouring
        in, 64
    skin mould, 63
Popping and pitting, 95—96
Potassium alum, 71
Pricking up, 15
Puddling in, 81
PVA, *see* Polyvinylacetate
PVC, *see* Polyvinylchloride

Rabbets, *see* Rebates
Radius rods, 27
Rapid-hardening cement, 4
Rebates, 48
    on fibrous plaster casts, 73,
        86, 87
Reinforcement, 72, 73
    see also Canvas, Wood lath
Render coat, 14–15
Repairing, 6, 102–107
Retarded hemihydrate plasters, 2
Retarders, 1, 71
Reveals, 24–25
Reverse moulds, 46–47, 56
Roof screeds, 40
Ropes, 76
Roughcast, 31
Rules,
    featheredge, 18, 20
    floating, 17, 18, 41
    joint, 94
Running, *see* Mouldings, run
Running moulds, 46 et seq.
    operation when running,
        56–59

Sand, 6–7
Scratch coats, 13, 15
Scratcher, 16
Screeds, 16, 18, 27, 40–42, 43
    curing, 44–45
Scrim, 35–36, 73
Seasoning,
    models, 64–65
    moulds for casting, 64, 65, 78
Seconds, 72
Sections, obtaining, 105
Setting coats, 18–20
Shellac, 51, 53–54, 78–79
Shelling, 80
Sirapite, 2, 9, 22
Size, 71

Skin moulds, 63
    as *in-situ* squeezes, 107
    backs for, 66–69
    PVC, pouring of, 66
Skirtings, coved, 42–43
Slipper, 46, 48, 50
Small tool, 93
Solid plastering, definition, 10
Spatterdash, 9, 14
Spinning, 59–62
Splash brush, 79
Spotboard and stand, 12
Square, 17
Squeeze moulds, 105–107
Staircases, 43–44
Stopping, 91–94
Storage,
    of materials, 8
    of plaster, 101
    of plaster casts, 84
    of plasterboards, 8
Strike-offs, 47, 73–75
Striking off, 80, 82
Suction,
    for dashing, 31
    for making good, 52, 92, 102
    for running, 55–56
    of a background, 13
    of a backing coat, 15
    of a waterproofed backing
        coat, 34

Templates, 28
Textured finishes, 33
Thermal insulation, 38–39
Thistle finish, 9
Transport of plaster casts, 84
Trowels,
    dashing/harling, 32
    gauging, 93
    internal angle, 21

Trowels (*cont.*)
    laying on, 10–12, 19
    skirting, 42–43
Tyrolean finish, 16, 30, 31

Vermiculite, 8, 9

Wads, 75

Walls, retaining, for open moulds,
    65, 67
Waterproof plastering, 33–34
Water-repellent cement, 4
Wet dashing, 31
Wood laths, 13
    boxed, 76
    in casts, 73, 76–78, 81
Woodwool slabs, 9

CARPENTRY AND JOINERY

A R WHITTICK

The basic aspects of the work of the carpenter and joiner are explained in a way that will appeal both to the handyman who wants to know how to maintain the joinery and timber structure of his home, and to the craft student.

CONTENTS: Hand and portable power tools. Timber and built-up materials. Timber defects, pests and diseases. Joints, fastenings and fixings. Temporary work. Carcassing. First and second fixing. Ironmongery and fittings. Basic woodworking machinery. Joinery. Preserving and finishing timber. Fences and gates. Repairs and maintenance. Index.

160 pages   1974   0 408 00375 8

Newnes Technical Books
Borough Green, Sevenoaks, Kent TN15 8PH

QUESTIONS & ANSWERS

CENTRAL HEATING

W H JOHNSON

This is a brief but comprehensive survey of domestic heating systems for both layman and student. The emphasis throughout is on fuel (and hence cost) saving, and how to achieve an acceptable standard of comfort for the least expenditure of energy.

CONTENTS: Insulation. Central heating systems. Wet systems. Boilers. Dry systems. The fuels. Domestic hot water. Controls. Some useful calculations. Some conclusions. Index.

128 pages   1980   0 408 00459 2

Newnes Technical Books
Borough Green, Sevenoaks, Kent TN15 8PH

QUESTIONS & ANSWERS

ELECTRIC WIRING

Henry A Miller

This book gives clear and logical guidance on electric wiring in its broadest sense – the installation of electrical equipment in buildings generally, including control and protection. Emphasising safety, it is based on the IEE Regulations for the Electrical Equipment of Buildings.

CONTENTS: Introduction. Protective requirements. Wiring systems. Lighting. Socket outlets, heating and cooking. Motors. Bells and alarm systems. Inspection and testing. Index.

128 pages   1974   0 408 00152 6

Newnes Technical Books
Borough Green, Sevenoaks, Kent TN15 8PH

QUESTIONS & ANSWERS

PAINTING AND DECORATING

A Fulcher and others

The authors, with many years' practical and teaching experience, provide a wealth of information on the materials, processes and methods of working required for decorating a house both inside and outside. The book will answer many DIY questions and also be a reference source for craft students.

CONTENTS: Tools and equipment. Preparation and painting of surfaces. Wallcoverings. Paint defects. Scaffolding. Work procedure. Index.

128 pages   1978   0 408 00321 9

Newnes Technical Books
Borough Green, Sevenoaks, Kent TN15 8PH

Questions and Answers

PLUMBING

A Johnson

This book gives concise and reliable information
on the principles of good plumbing design and
on the use of plumbing materials both old and
new. Guidance on calculations is included,
and there is useful tabulated information.

CONTENTS: Pipes and pipework. Water supply.
Sanitary appliances. Drainage. Above ground
drainage. Domestic hot water supply. Sheet
weathering. Calculations. Appendix. Index.

96 pages   1974   0 408 00136 4

Newnes Technical Books
Borough Green, Sevenoaks, Kent TN15 8PH